SECRET
AGENTS

THE MENACE
OF
EMERGING
INFECTIONS

Madeline Drexler

Joseph Henry Press
Washington, D.C.

JOSEPH HENRY PRESS • **2101 Constitution Avenue, N.W.** • **Washington, D.C. 20418**

The Joseph Henry Press, an imprint of the National Academy Press, was created with the goal of making books on science, technology, and health more widely available to professionals and the public. Joseph Henry was one of the founders of the National Academy of Sciences and a leader in early American science.

Library of Congress Cataloging-in-Publication Data

Drexler, Madeline, 1954-
  Secret agents : the menace of emerging infections / Madeline
Drexler.
      p. cm.
Includes bibliographical references and index.
  ISBN 0-309-07638-2
  1. Epidemiology—Popular works. 2. Communicable
diseases—United States. I. Title.
  RA653 .D74 2002
  614.4—dc21

2001007832

Cover design by Bill Stanton. Cover photography by David Oliver/Getty Images and Stephen Mautner.

Illustrations by Kathryn Born.

Contents

Acknowledgments

Book writing is a group activity. I am grateful to many people, both new acquaintances and old friends, for sharing their expertise and their enthusiasm.

Thank you to the scientists and physicians who reviewed chapters and offered valuable comments: Steve Brickner, Jim Hughes, Tom Inglesby, Jim LeDuc, David Morens, Mike Osterholm, and Bob Shope. During the course of researching this subject, I was privileged to have wide-ranging conversations with Fred Angulo, Bill Foege, Patty Griffin, Duane Gubler, Peggy Hamburg, D. A. Henderson, Donald Hopkins, Jim Hughes, Marci Layton, Tracey McNamara, David Morens, Steve Morse, Mike Osterholm, Tara O'Toole, C. J. Peters, David Relman, Jeff Taubenberger, and Rob Tauxe. I am also deeply obliged to the nearly 250 researchers who granted me interviews.

I am especially beholden to Joshua Lederberg, who chaired the Institute of Medicine's Forum on Emerging Infections, for his visionary leadership on this issue. Thank you to Ken Shine, president of the Institute of Medicine, for championing this book. And thanks to Jonathan Davis, who made me welcome at the Forum's workshops.

My gratitude goes to the Eli Lilly and Company Foundation, for their generous sponsorship of this project, and to Gail Cassell, for her persistence and dedication in arranging this support. At the National Museum of Natural History, ornithologist Carla Dove and her colleagues saved the day with a bird-in-hand.

I owe a great deal to my colleagues at Joseph Henry Press: my editor Stephen Mautner, for inviting me to write about this fascinating topic and for his unwavering confidence in me; Ann Merchant and Robin Pinnel, for getting out the word; and all the staff who took so much care with this book.

Caroline DeFilippo was my whirlwind research assistant. While completing her undergraduate degree at Harvard University, she somehow found the time to scour the school's vast libraries for the most arcane references. I look forward to watching Caroline's own career in public health unfold, and to catching up together over Bartley's hamburgers-and-fries. During the summer of 1999, I also received research help from Harvard student Navaz Karanjia. Thank you to Sharon Crandall and Phyllis Strimling at the Radcliffe Research Partnership, for underwriting this assistance.

For practical advice and on-call cheerleading and for putting up with my authorial anxieties, I am indebted to Suzanne Berne, Cynthia Enloe, the Graff family, Ann Parson, Joni Seager, Joe Shay, Seth Shulman, and Pat Thomas. Thank you to Marcia Bartusiak, for matching author and editor. And thanks to Kevin Shay, who after a desperate email plea quickly came up with the title. My mother, Barbara Drexler, and my siblings, Michael Drexler and Suzanne Drexler, were always behind me. My dear father, Milton Drexler, passed away during the early phases of this project, but I think he would have slipped this volume in the stack of histories and biographies he always had going.

I was lucky to have a superb editorial team. My dear friend and writing buddy Laura Zimmerman read each chapter several times, always with a thoughtful and penetrating analysis, always making me answer the question, Who cares? For the last time, let me say to Laura: I couldn't have done it without you. Chris Jerome deftly edited the final manuscript, and gave me warm encouragement when my spirits were flagging.

My most beloved critic, E. J. Graff, not only edited every draft of every chapter until I got it right, but cooked the soup and rented the videos and arranged our Saturday nights. I thank E. J. for believing in me and for making my life better in every way.

January 2002

SECRET
AGENTS

Chapter 1

Disease in Disguise

Infection is an inescapable part of life. All creatures feast on other creatures and in turn are feasted upon, in a kind of Escheresque food chain. When humans are the meal, we call it infectious disease. This book is about today's new and emerging infections—those that have increased in attack rate or geographic range, or threaten to do so. It explores why these infections are materializing now and why they will never go away. And it tells the stories of scientists racing to catch up with invisible adversaries superior in speed and guile. Each chapter will look at a different threat: animal- and insectborne diseases, foodborne pathogens, antibiotic resistance, pandemic influenza, infectious causes of chronic disease, and bioterrorism.

The list is not unique. Any slice of time yields a catalog of infectious disease, new and old. The mid-1970s to the early 1980s were especially fecund. First came the swine flu debacle of February 1976, when the U.S. government pulled out all stops preparing for a repeat—which never occurred—of the deadly 1918 influenza pandemic. Summer of 1976 brought a bona fide novelty, Legionnaires' disease, which struck in Philadelphia, killed 34 people, and stumped the nation's top infectious disease experts for nearly six months. As other outbreaks followed in short order, a new vocabulary arose: Lyme disease, toxic shock syndrome, *E. coli* O157:H7, STDs, Ebola virus. In June 1981 came the now-landmark report from the Centers for Disease Control and Prevention (CDC). Nine brief paragraphs described a strange cluster of fatal symptoms among five gay men in Los Angeles: acquired immunodeficiency syndrome, or AIDS, which had already silently infected 250,000 in the United States.

The same tricks of biology and convergence of habits and slips of surveillance that brought us AIDS are constantly drawing agents into our midst. As the world was grimly marking the twentieth anniversary of that CDC report, other infections were quietly exacting a toll. In the upper Midwest, a commonplace staph infection had turned inexplicably virulent, impervious to one of the most powerful drugs used to treat it; four healthy children and a strapping college student suddenly died, and hundreds were infected. *E. coli* killed a little girl who indulged in watermelon slices at a Sizzler restaurant in Milwaukee, one of 5,000 annual deaths from foodborne pathogens. Mosquitoes and birds carried West Nile virus from New York City to the Deep South and beyond. Nature lovers and suburban homeowners were catching tickborne Lyme disease. In the terror-filled autumn of 2001, deliberately planted anthrax spores killed and infected Americans, and U.S. public health officials girded for the return of smallpox.

As bacteriologist and historian Hans Zinsser wrote in 1934, "However secure and well-regulated civilized life may become, bacteria, Protozoa, viruses, infected fleas, lice, ticks, mosquitoes, and bedbugs will always lurk in the shadows ready to pounce when neglect, poverty, famine, or war lets down the defenses. And even in normal times they prey on the weak, the very young and the very old, living along with us, in mysterious obscurity waiting their opportunities."

Through media coverage and sometimes sensationalistic reportage and fiction, Americans have awakened to the emerging infection threat. But our anxieties often focus on the trivial. The CDC has had to contend with bogus reports of imported bananas carrying flesh-eating bacteria, drug addicts placing HIV-infected needles in pay phone coin-return boxes, virus-soaked sponges arriving with the mail. Late in 2001, the rumors were of war—germ war. Perhaps because we increasingly feel at the mercy of a world out of control, bizarre fictional health threats quickly take root. But these fantasies mask more likely—and commonplace—sources of danger.

To offer a local view of a universal and authentic peril, this book focuses on the United States. In truth, the distinction between national and global threats is artificial. Infectious agents need no visas. Secret agents shadow ecological change everywhere, and the pace of change is speeding up. U.S. health officials scan the terrain beyond our national borders in part because many of those infections can be transported, in people or goods or animals, to this country. Scientists must also keep apprised of pathogens abroad because related organisms may already be here, lying in wait. When a deadly pulmonary syndrome turned up for the first time in the Southwest in 1993, for example, researchers were at first baffled—until diagnostic tests developed decades earlier by the Defense Department, to identify a related virus in Korea that caused a very different pattern of disease, provided the first laboratory evidence that they were dealing with the genus Hantavirus.

What we face today is what we have always faced: one plague (the word comes from the Greek for "blow" or "strike") after another, and usually several at once. A 2000 report from U.S. intelligence agencies described infectious diseases as a "nontraditional" threat. But history shows emerging infections to be the *most* traditional threat, both to individuals and to nations.

"Some mysterious, intangible thing ..."

Before the recent headline accounts of avian flu, a silent hepatitis C epidemic, parasite-contaminated water slides, new variant

Creutzfeldt-Jakob disease, and anthrax, we had smallpox, typhus, polio, cholera, rabies, and the Black Death. We still have them. None have been consigned to history; in some cases, they are lethally intractable today. Even grisly reports of flesh-eating bacteria are nothing new. In the fifth century BC, Hippocrates described a condition in which "Flesh, sinews and bones fell away in large quantities . . . There were many deaths. The course of the disease was the same to whatever part of the body it spread."

Small human populations suffer the oldest diseases of humankind: those that are either chronic, such as leprosy or herpes, or that have reservoirs in animals or the soil, such as yellow fever, the virus of which circulates in monkeys. Only when a community is dense and filthy enough to keep spreading germs and big enough to keep supplying new susceptibles do such infections as measles, smallpox, typhoid, and influenza—crowd diseases, or "zymotics"—stay in circulation.

The shattering "plagues" of history (some of which actually were plague) bespoke *Homo sapiens*'s immunological naiveté. In AD 79, an outbreak believed to be malaria contributed to the Roman Empire's fall. The plague of Antoninus (AD 166–180), probably smallpox, killed between one-fourth and one-third of Italy's population. The plague of Justinian (AD 542–543), one of the first documented cases of ratborne bubonic plague, was "a pestilence by which the whole human race came near to being annihilated," in the words of Byzantine historian Procopius, killing 10,000 people daily in Constantinople and eventually spreading as far north as Denmark.

Over the next centuries, wave upon wave of emerging infections shaped human history. Between 1346 and 1350, one-third of Europe's population died of bubonic plague, spread from Asia to Europe by Mongol armies, whose retinues of rodents carried infected fleas that subsequently bit humans. In the sixteenth and seventeenth centuries, slave ships from West Africa brought yellow fever and its mosquito vector, *Aedes aegypti*, to the New World. Smallpox, transported to the Americas by Spanish conquistadors, killed one-third of the relatively disease-free native population, and was followed by a similarly lethal

onslaught of measles. Diseases introduced from Europe killed an estimated 95 percent of the pre-Columbian Native American population.
To a far lesser extent, exotic infections attacked in the other direction
too. European colonials succumbed to malaria, yellow fever, and other
endemic infections in tropical Africa, India, Southeast Asia, and New
Guinea. Syphilis, a bacterial scourge that relies on a mobile population
and indiscriminate sexual contact, appeared in 1494—perhaps with
Columbus's returning soldiers.

The Industrial Revolution of the nineteenth century amplified such
diseases as tuberculosis, an ancient bacterial infection that thrives in
squalid close quarters. Illnesses from contaminated food and water,
such as typhoid and cholera, also went on a spree. Nevertheless, life
expectancy rose during the nineteenth century—perhaps, as some say,
because a biological equilibrium between our bodies and ever-present
pathogens was achieved. The twentieth-century British physician and
historian Thomas McKeown argued that the improvement stemmed
from better nutrition, which built stronger resistance, and from sanitation measures such as sewage disposal, which reduced our exposure to
disease-causing microbes. More recently, scholars have suggested that
personal health practices, such as boiling drinking water and isolating
sick relatives, may have made some difference.

Beginning in the late 1870s, the new discipline of bacteriology
found the agents that caused cholera, tuberculosis (TB), gonorrhea,
typhoid, and scarlet fever. "We no longer grope after some mysterious,
intangible thing, before which we must bow down or burn something,
as if it were some demon which we would exorcize," wrote a physician
in 1890. By 1900 scientists widely agreed that microorganisms—spread
by casual contact, food and water contamination, insects, and even (in
the cases of typhoid and TB) healthy human carriers—caused communicable diseases. These discoveries spurred expansion of government
health initiatives such as water purification, food inspection, and rodent control, as well as more awareness of individual hygiene measures
such as covering a cough or washing hands before eating. The years
1890 to 1930 were seen as the golden era of the American public health

movement. As a 1924 *Scientific American* editorial put it, "the natural outcome of the struggle between mankind and microbe has always favored man."

The Price of Complacency

Medicine soon targeted the discrete organisms behind infectious disease. In the 1930s and 1940s came specific antimicrobial therapies such as sulfa compounds and penicillin. By the mid-1960s, hundreds of new antibiotics were available to treat such afflictions as gonorrhea, syphilis, pneumonia, TB, bacterial meningitis, typhoid fever, even bubonic plague, while new vaccines prevented epidemics of measles, rubella, and polio. The study of infectious disease became unfashionable. Ambitious young doctors were advised not to specialize in it. In his classic 1962 text *The Natural History of Infectious Disease*, the Australian Nobel laureate Sir MacFarlane Burnet mused that "to write about infectious disease is almost to write of something that has passed into history."

Between 1970 and 1975, as the National Institutes of Health budget increased 100 percent, the budget for its infectious disease arm rose only 30 percent. Cancer and heart disease seemed the last unconquered frontiers. Public health officials assumed that proper hygiene, sanitation, vaccination, antibiotic use, and good hospital care would keep the bugs at bay. The *New Yorker* medical writer Berton Roueché could afford a tone of amused detachment; infectious disease was ripe for elegy.

The belief that enlightened policies would finally wipe out infections peaked in 1978, when United Nations members signed the Health for All 2000 accord, which predicted that even the most cash-strapped nations would undergo a health transition before the millennium. That assumption was rooted in the public health concept of the "epidemiological transition": as nations develop, life expectancy increases and deaths due to pestilential infections give way to deaths caused by heart disease, malignancies, and other chronic afflictions of old age. Now, at the beginning of the twenty-first century, the notion of a neat "before" and "after" pattern of disease tethered to "progress" seems hollow. "A hundred years from now," suggests Columbia University virologist Steve Morse, "we may still be worried about tuberculosis while having managed to get rid of most of the underlying causes of heart disease. That's really turning the conventional wisdom on its head. Could you imagine a well-fed society that has got all the right interventions—even a little nanotechnology to clean out the arteries—and another HIV comes along?"

When HIV came along for the first time, government leaders were looking the other way; in the early 1980s, too, public health agencies had stopped surveillance for drug-resistant tuberculosis just as the disease started to rise (it soon became endemic in New York City, bred by poverty, a surge of immigration, urban crowding, and the concomitant AIDS epidemic). "The national public health capacity was allowed to lapse during the eighties and early nineties," says Jim Hughes, director of the CDC's National Center for Infectious Diseases. "We're still scrambling and playing catch-up." That indifference translated into numbers. In the two decades after 1980, the U.S. infectious disease mortality rate jumped 58 percent; even after removing AIDS from the tally during that period, the United States saw a 22 percent rise in infectious disease deaths. Public health requires long-range planning and strong commitment demonstrated by steady funding, because surveillance and response focus on improbable biological events that happen inevitably, though unpredictably. In 1998, the CDC issued a detailed five-year plan to prepare the nation for emerging infections, but congressional funding has lagged. In fiscal year 2001, though the agency estimated it would need $260 million, it received only $139 million.

A Millennium in a Fortnight

The microbiologist René Dubos described life itself as "emergence," a perpetual process of evolution and adaptation. In this sense, emerging infections are merely chapters in a Darwinian saga stretching over eons. Conditions ceaselessly shift—temperatures rise and fall, habitats flourish and perish, food supplies grow lush and dwindle—and all living things transform themselves or die.

In the perpetual drama of emerging infections, nature's undercover operatives are the same: viruses, bacteria, fungi, protozoa—a crew collectively known as "microorganisms" (colloquially, "microbe" refers only to bacteria). Nor has the plot changed. Just as all mystery writing draws on five or six plotlines, so biologists have found only a handful of pathogenic strategies, breathtakingly elaborated.

Microorganisms play the survival game exceedingly well. For one thing, they adapt far more quickly than do humans to the scene shifts around them. Humans crank out a new generation every 20 years or so; bacteria do it every 20 to 30 minutes, and viruses even faster. Richard Krause, former director of the National Institute of Allergy and Infectious Diseases (NIAID), calls this microbial pace "a millennium in a fortnight." Natural selection—the process by which genetically better-adapted individuals leave more progeny and thus transmit those desirable characteristics—operates far more efficiently in the microbial world. And because they assemble in enormous numbers, viruses and bacteria can support considerable variety in their communities, including the mutated oddballs that may shine when circumstances change. A billion bacteria inhabiting a thimble can be virtually wiped out on Monday and be back in full force by Tuesday.

"When you look at the relationship between bugs and humans, the more important thing to look at is the bug," says NIAID medical epidemiologist David Morens. "When an enterovirus like polio goes through the human gastrointestinal tract in three days, its genome mutates about two percent. That level of mutation—two percent of the genome—has taken the human species eight million years to accomplish. So who's going to adapt to whom?" Pitted against such nimble competition, the human capacity to evolve "may be dismissed as almost totally inconsequential," adds Joshua Lederberg, professor emeritus at Rockefeller University, whose discoveries about genetic recombination in bacteria earned him the Nobel Prize in 1958.

With microorganisms, less is more. Consider their simplicity. A virus—the word comes from a Latin term for "poisonous substance"—is nothing more than nucleic acid, DNA or RNA, surrounded by a shell made of protein and sometimes also of fatty materials called lipids. The biologist Peter Medawar once observed that a virus is "a piece of bad news wrapped in protein." Viruses are tiny, ranging in size from about 20 to 400 nanometers in diameter—millions can fit in the period at the end of this sentence—and visible only through an electron microscope. Some are rod-shaped, others round and 20-sided, and yet others have fanciful forms with multisided "heads" and cylindrical "tails." Outside a living cell, a virus is a dormant particle, lacking the raw materials for synthesis. Only when it enters a congenial host cell does it explode into action, hijacking the cell's metabolic machinery to produce copies of itself that may burst out of infected cells or simply bud off a cell membrane. This squatter's existence means that viruses cannot be cultured in artificial media; they can only be propagated in live cells, fertilized eggs, tissue cultures, or bacteria. Scientists believe that viruses were originally fragments of genetic material derived from cellular organisms—perhaps bare nucleic acid, or pieces of DNA from bacteria or higher animals.

Viruses make us sick by killing host cells or by skewing the cells' function. Our bodies often respond with fever (heat inactivates many viruses), the secretion of a chemical called interferon (which blocks virus replication), or by marshaling the immune system's antibodies and other cells targeted to the invader. But viral infections are hard to fight once the process is under way because our immune responses usually kick in too late to subdue them. The virus that causes AIDS is especially clever, targeting the immune system itself. And because viruses commandeer the machinery of living cells to replicate, it's been hard to develop drugs that combat the infection without also harming our cells. Only a small number of antivirals are available for particular infections. Doctors generally treat viral infections by trying to ease specific symptoms, such as fever or dehydration. In other words, although their symptoms can be treated, viruses themselves are so far undefeated in their ongoing war with human medicine.

Bacteria are about 1,000 times larger than viruses and more self-sufficient. One-celled organisms that are generally visible under a light microscope, bacteria are known as "prokaryotic"—so primitive, they lack a membrane-bound nucleus with neatly linear chromosomes inside. Instead, bacteria usually carry a tangled necklace of DNA joined at the ends, and sometimes smaller rings of DNA known as plasmids, which contain genes that enable a bacterium to manufacture proteins beyond its usual repertoire. Unlike more advanced organisms, bacteria carry one set of chromosomes instead of two, an arrangement that means that every gene counts and every selective advantage must be conserved. By taking in new genetic material instead of slowly adapting over millions of years, they evolve in quantum bursts. Bacteria come in three name-conferring shapes: spherical (coccus), rodlike (bacillus), and curved (vibrio, spirillum, or spirochete). Reputed to be the most ancient organisms, imprinted in fossils more than three billion years old, bacteria have evolved a vast range of behaviors over a vast range of habitats.

Over eons, bacteria have learned tricks to help them cleave to cells, make paralyzing poisons, elude or suppress our bodies' defenses, and shrug off drugs and antibodies. They can pick up genes from almost everywhere: from other bacteria, from viruses, from plants, even from yeasts, in a round-the-clock flea market that's a source of impressive plasticity. The pneumococcus bacterium, which causes pneumonia and other ills, soaks up DNA that has spilled out of dead or dying brethren—inheritance by cannibalism that has imparted, among other things, antibiotic resistance. When a virus picked up a toxin gene from a deadly *Shigella dysenteriae* and inserted it into a harmless *E. coli*, it created *E. coli* O157:H7, a mean bacterial hybrid that clings to mucosal surfaces in the human intestine and produces toxins that trigger hemolytic uremic syndrome, the most common cause of acute kidney failure in children. Such mixing-and-matching is "an exquisite example of genetic engineering carried out in the wild," according to Stanford University molecular biologist Lucy Shapiro. Bacteria even employ a stealth tactic known as phase variation, hiding their immunity-provoking surface proteins and sugars to fool the body's defenses. According to Joshua Lederberg, "That says they've got a memory. They're carrying about pieces of their evolutionary history in unexpressed forms, waiting to be expressed."

Bacteria inflict damage in a different way than viruses. Sometimes they multiply so rapidly they crowd out host tissues and disrupt normal function. Sometimes they kill cells and tissues outright. Sometimes they manufacture toxins that can paralyze, destroy metabolic pathways, or generate a massive immune reaction that is itself toxic. Drug-resistant bacteria often make an enzyme that destroys antibiotics or spits them out. Special virulence factors enable bacteria to penetrate cells, gather nutrients for growth and survival, and evade the host's defenses by jamming or slipping in under the immune system's radar. Furthermore, bacteria don't attack until their numbers are high enough to establish an infection. This wireless communication system, called "quorum sensing," enables microbes to coordinate their activities. The bacteria that congregate in the slimy biofilms such as dental plaque actually assume individual specialized tasks. Despite these feints and stratagems, bacteria remain easier to treat than viruses. Because they are free-living, and because their structure differs from that of mammalian cells, they are more susceptible to drugs delivered via the bloodstream.

A more mysterious class of infectious agents—the newly discovered triggers of animal and human brain diseases such as bovine spongiform encephalopathy (BSE, or mad cow disease) and its human counterpart, new variant Creutzfeldt-Jakob disease—apparently repeal the laws of biology. Called prions (pronounced *pre´-ons*), these proteins are folded in an abnormal way; when they come in contact with other proteins, they turn *them* into prions, setting off a chain reaction that eventually riddles the brain with holes. A cow can contract BSE by eating one gram of prion-infected tissue—the size of a peppercorn—from another cow. Unlike viruses or bacteria, prions can't reproduce and evoke no immune response. More frightening, they resist heat, ultraviolet light, radiation, and sterilization.

The traditional wisdom about emerging pathogens is that they are noxious because they are new—that is, ill adapted to a human host. Indeed, their destructive power is supposedly a tip-off to their recent arrival and to a bad biological fit. Animal viruses, such as the Ebola virus or the Sin Nombre virus that causes hantavirus pulmonary syndrome, can trigger lavish and bizarre symptoms—from hemorrhagic breakdown to acute respiratory failure—because our immune response has not evolved with the virus. Over the long haul, microorganisms and humans usually reach a subtle accommodation. Humans acquire resistance to the infectious agent while the parasite becomes milder, permitting us to survive its assault and permitting it to transmit its genes to someone else. Microorganisms *need* their hosts to survive; a dead host is a dead-end. The reason the lethal spore-forming bacillus *Clostridium botulinum*—the cause of botulism—hasn't leveled our species is because when it kills us with its toxins, it kills its prospects for spread.

In unlocking microbes' secrets, scientists have gained respect for these hidden adversaries. Humans, they agree, are an encumbered species, composed of too many working parts, fragile when stressed. Bacteria, having evolved for more than three billion years, are supremely well adapted and elegantly stripped down. "I am in awe of these minute creatures," says Stanford University microbiologist Stanley Falkow. "They know more about the biology of the human cell than most cell biologists. They know how to tweak it and how to exploit it."

Viruses may be more clever yet. "Some viruses are enormous, and have large numbers of genes that do a lot of things. But other viruses are streamlined, and those are the ones that I really admire," says molecular biologist and composer Jeffery Taubenberger, who has studied the 1918 influenza virus. "These viruses are so tightly packaged that several different genes can be coded from the same sequence by starting at different places—the sequences actually overlap to conserve space. Genetically, they work like a Bach fugue. In a purely biological sense, they're incredibly beautiful."

Traffic Patterns

As far back as the second century AD, in one of his philosophical *Table Talks*, the historian Plutarch argued that new classes of disease arise because of profound changes in the way we live. This is no less true today. If you want to know where emerging infections come from and how they gain a foothold, take a look at your own life. Virtually every aspect of American culture—from where we live to where we play, from how we raise livestock to how we raise children—is changing. Change creates new markets, you might say, for pathogens. And these agents have a knack for leveraging the slimmest advantage.

The story of emerging infections is like the story of a dangerous intersection that needs a stop sign. Every time two cars collide, bystanders shake their heads and wonder why it doesn't happen more often. In the case of emerging infections, the collision is between pathogens and people. These crashes take place partly because there's so much snarling traffic—not only of humans, but of animals, plants, seeds, insects, and all manner of life. Every day, more than two million people worldwide cross national borders. Every year, more than 1.5 billion people travel by air. The United States alone hosts 47 million visitors yearly. "In the old days, our neighbors were Canada and Mexico," says CDC director Jeffrey Koplan. "Nowadays, with the frequency and speed of air travel, our neighbors are Sri Lanka and Paraguay and you name it."

Just as, in the nineteenth century, cholera traveled on steamships to Europe and Africa, so in the early 1990s cholera reached the oyster beds of Mobile Bay by stowing away in the bilgewater of ships from Latin America. Though trucks, freighters, and airplanes may have replaced caravans and steamships, the results are the same. Look in your shopping cart. Today's international cornucopia has included *Cyclospora*-tainted raspberries from Guatemala, *Shigella*-contaminated green onions from Mexico, *Salmonella*-saturated alfalfa sprouts from the Netherlands, *E. coli*-covered carrots from Peru, and other invisibly poisoned fare. Mike Osterholm, former Minnesota state epidemiologist and an esteemed disease sleuth, describes the clinical sequelae as "classic traveler's diarrhea for individuals who never leave home."

Any pathogen, not just those present in food, can be virtually anywhere within 48 hours. In 1998, a Ukrainian émigré on a Paris-to-New York airplane flight infected 13 other passengers with drug-resistant tuberculosis. In 2001, an Ontario hospital went on red alert when a Congolese woman arrived with what looked to be a hemorrhagic virus such as Ebola. Though she was eventually diagnosed with malaria, public health officials were once again reminded of the ticketless travel arrangements of exotic pathogens. The scenario they most dreaded was a brand-new disease that couldn't be diagnosed.

Microbes also mix it up when we settle down—especially at the margins of the wild. Lyme disease came to us courtesy of nineteenth-century deforestation in the Northeast, followed by patchy and less diversified second-growth forests. These nurtured deer populations that, without natural predators, exploded, coinciding with an exurban surge toward the idyllic fringes of "nature." Coastal population growth has led to contamination of shellfish beds with human waste, fostering the transmission of viral and bacterial pathogens. Human encroachment on the tropical rainforest may open the way for hemorrhagic fever viruses and perhaps even HIV's mysterious retroviral cousins; in Africa, HIV-like organisms have turned up in more than a dozen species of primates hunted for bush meat or kept as pets.

But communing with nature isn't the only path to pathogens: microbes also love crowds. In 1900, only 5 percent of the population lived in cities of over 100,000 residents. By the year 2025, 65 percent of the population in developing regions will inhabit cities. Dense urban enclaves will be simultaneously magnets for infections from isolated rural areas and launch pads that allow pathogens to reach other fast-growing populations. Overwhelmed by unsafe water, poor sanitation, and widespread poverty, tomorrow's megacities will become cauldrons for new infections.

The devastating 1998–1999 Nipah virus outbreak in Malaysia that killed nearly a third of infected people probably sprang from intensive pig raising, which permitted a novel virus (probably carried by fruit bats) to propagate and then jump to farmers. Pig farms acted as megacities for the deadly agent, and pig farmers became sentinel cases. In industrialized countries, day care centers are notably noxious settings in which the combination of frequent infections, susceptible children, poor hygiene, and high antimicrobial use breeds diarrheal diseases and antibiotic-resistant microbes such as *Streptococcus pneumoniae,* a common cause of ear infections and pneumonia.

Keeping ourselves alive longer also, paradoxically, increases our susceptibility to infection. In 1900 only 4 percent of the U.S. population was over 65; in 2040 it will reach almost 25 percent. Elderly people, with their fading immunity, are at the mercy of microorganisms that are normally benign. While chemotherapy and other immunosuppressive treatments have enabled people to live with cancer and other illnesses, they also increase our susceptibility to such ubiquitous pathogens as cytomegalovirus, which can cause pneumonia and eye infections, as well as to new threats such as West Nile virus.

Modern technologies intended to make our lives easier may also

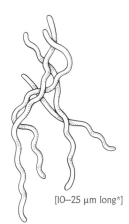

[10–25 μm long*]

Borrelia burgdorferi
Agent of Lyme disease

---

*Bacteria and viruses are almost unimaginably tiny. Throughout this book, illustrations of microorganisms will include their approximate dimensions. Bacteria are usually measured in microns (one micron equals one millionth of a meter), while viruses are measured in the vanishingly minuscule unit of nanometers (one nanometer equals one billionth of a meter). To give a sense of these measures, consider that the width of a single human hair is about 80 microns, or 80,000 nanometers—equal to approximately 2,000 copies of the West Nile virus. The abbreviation for measurements in microns is μm and for measurements in nanometers is nm.

make life easier for microbes. Take Legionnaires' disease. The bacterium, normally a habitué of moist soil and lakes, not only thrives in water of a narrowly warm temperature range, but also must be misted into tiny particles in order to penetrate deep into human lungs—and so is neatly accommodated by cooling towers, whirlpool spas, and even, according to a recent news account, the hot-water pipes in the Queen's shower at Buckingham Palace. More futuristically, organ transplants from pigs—contemplated as treatments for diabetes and Parkinson's disease, among other illnesses—could infect humans with porcine endogenous retroviruses, or PERVs, which are in the same class as HIV; critics fear that such viruses, which are insinuated in the donor pig's DNA, could interact with human viruses to create new, potentially dangerous species that might spread to the general population.

Bugs themselves are changing, often with a human assist. In 1954 the United States produced two million pounds of antibiotics. Today, it makes tens of millions of pounds per year, half or more administered to livestock. As a result, 70 percent of bacteria that cause the infections patients acquire in hospitals are resistant to at least one antibiotic, and the animals we eat have become factories for drug-resistant microbes. Our newest foe in the antibiotic wars—vancomycin-intermediate *Staphylococcus aureus*, or VISA—defies medicine's most powerful antibiotic; should VISA infections become common (five have been reported so far in the United States), simple scrapes could become mortal wounds, and surger ld be as dangerous as it was a century ago—a prompted some scientists to speak of a "post-us seventeen years to develop an antibiotic," t a bacterium can develop resistance virtually e putting our best players on the field, but the while their side has an endless supply of new players."

Finally, microorganisms may have a much broader reach than we have understood to this point. New DNA-based methods have revealed infectious agents behind diseases that in the past were blamed on lifestyle or the environment. Researchers have reason to be looking for infectious elements in atherosclerosis, Type I diabetes, multiple sclerosis, Crohn's disease, Guillain-Barré syndrome, Tourette's syndrome, even kidney stones. Someday, your annual physical may include a quick survey of "endogenous microflora" in the mouth, on the skin, and in the bowel, to predict current and future medical problems.

Using equally sophisticated epidemiological tools, could science pick up the next HIV before it spreads? The CDC is betting on it, with a project that scrutinizes sudden unexplained deaths and life-threatening illnesses among previously healthy people. The centers hope to pick up odd, sporadic infections and piece them together like a jigsaw puzzle before the illnesses spiral into public health emergencies.

***

In recent decades, our attention to infectious disease has pulsed on and off, following the signals of government interest and political fortitude. But for bugs, the beat goes on. "Let your guard down for a minute," says Joe McDade, deputy director of the National Center for Infectious Disease, "and the microbes are still there."

What many biologists fear most is a new deadly virus. Viruses are harder to fight with drugs, intimately entangled as they are with the genes and metabolic machinery in our cells. Viruses also seem to stimulate our immune systems more violently and self-destructively than do bacteria. And unlike with bacteria, which have known virulence factors, it's harder to predict whether a particular virus will radiate quickly or will be especially savage. "If you wanted to think of an Andromeda strain, you would think of, say, a virus with a short incubation period that's rapidly transmitted from person to person through aerosol, with potential for a high fatality rate and which could move very rapidly," says McDade, who in 1976 discovered the cause of Legionnaires' disease. Influenza currently fits the bill—but so could something else, such as a hypothetically mutated Ebola virus that spreads through the air.

Back during the Great Depression, when life seemed at a standstill, Hans Zinsser wrote that "Infectious disease is one of the few genuine adventures left in the world." Modern adventurers like to up the ante, but even the most extreme sports wouldn't produce the adrenaline of a race against pandemic influenza or a cloud of anthrax at the Super Bowl. In the field of infectious disease, reality is stranger than anything a writer could dream up. The most menacing bioterrorist is Mother Nature herself.

Chapter 2

Winged Victories

I n New York City, late August is a seasonal point of suspension
between summer languor and autumn snap, when nothing is supposed to happen. In August 1999, Marci Layton, the assistant com-
missioner for the Bureau of Communicable Disease in the New York
City Department of Health, was, like many residents, planning to flee
the city—to hike in the northern Adirondacks, the only wilderness
nearby where you can actually reach the top of a mountain and gaze
out for miles. If Layton relishes a long view, it's probably because she
directs about 100 outbreak investigations a year and closely monitors
surveillance for the 52 infectious diseases that the city requires doctors
to report. During the five days left before her weekend hike, she faced
the usual crises. A rare strain of *Salmonella* had struck more than 20
toddlers across the city, and Layton suspected a contaminated food.

Drug-resistant pneumococci hit three nursing homes. An untreatable strain of *Acinetobacter* was spreading across the city's intensive care units. Layton's department was winding down a campaign to vaccinate gay men against hepatitis A. And looming just three weeks away was a bioterrorism preparedness field exercise in the Bronx, to be dubbed City Safe—part field exercise, part educational extravaganza for the media, with simulated emergencies at a hospital, an airport, even the Commonwealth Edison plant.

But on Monday, August 23, Layton's phone rang, and her life—not to mention her weekend plans—changed. On the line was Deborah Asnis, an infectious disease doctor at Flushing Hospital Medical Center in northern Queens. Two cases troubled Asnis. One was a 60-year-old man who had entered the hospital on August 12, weak from days of vomiting and diarrhea. Rather than quickly rebounding with intravenous fluids, he slid downhill. He had become confused and then developed what doctors call flaccid paralysis of unknown etiology of the upper and lower extremities: his limbs froze, and nobody knew why. Transferred to the intensive care unit, he had been treated as if he had Guillain-Barré syndrome, an inflammation of the nerves, but none of the conventional treatments had worked. Asnis was also worried about a 75-year-old man in her care. He had been wheeled in with a high fever, shaking uncontrollably. Like the other patient, he too became extraordinarily weak over the next few days, so weak he couldn't move his arms or legs. Both men had spinal fluid with high levels of white blood cells and protein, which hinted at a viral encephalitis. But encephalitis doesn't usually cause muscle weakness. And since one of the differential diagnoses for Guillain-Barré is botulism—one of the diseases, in this case caused by food toxins, that doctors must report—Asnis dialed the health department.

Marci Layton gets this sort of call at least once a week. An infectious disease specialist, she had chosen the field in order to satisfy a taste for exotic locales. Layton had volunteered in a clinic outside Katmandu, Nepal. She had worked in a hospital in Thailand. She had covered for a doctor in a 12-bed hospital in northern Alaska, flying by helicopter to a clinic on the north slope. When a CDC employee told her she could find the same range of infectious exotica in New York City and still sleep in her own bed at night, she signed on in 1992 as the agency's epidemic intelligence officer there, later joining the city health department. Slight and of middling height, with curly red hair parted down the middle, she has a way of somewhat shyly looking up with steady blue eyes as she speaks, which is usually at a brisk clip. Unlike many physicians in public health, she had actually enjoyed taking care of patients when she specialized in infectious diseases, and still welcomed the chance to puzzle through a diagnosis.

In one respect, though, this afternoon's call was unusual. Flushing Hospital Medical Center had been in the news for years. A small fish in a big tank of recent hospital mergers, it had entered into a disastrous alliance with another institution in order to survive. In 1998, Flushing Hospital had filed for bankruptcy. So badly had the facility been tarred in the media, many people in the neighborhood thought it had already been shut down. When Asnis later told her colleagues that she had called the health department for advice, they were aghast. "When we talk to the health department," she explained, "they're usually coming in and condemning the place."

Asnis, however, was unfazed. Forty-three years old, she had come from a medical family. Her father had been a dentist, and two older brothers became doctors. She married a doctor. Her mother had strongly encouraged her to enter the field. "She wanted me to do anything it took to be independent," said Asnis, a small, striking woman with short dark hair, lively brown eyes, and bright red lipstick and nail polish. Asnis had herself grown up in Whitestone—the same patch of northern Queens, bracketed by the Bronx-Whitestone and Throgs Neck bridges, from which her patients came—before her family moved out to Long Island. She settled on her specialty because, as she put it, "All the enigmas and fascinomas were in infectious disease."

During the phone conversation, Layton quickly ruled out botulism—the symptoms and the lab findings didn't fit. She urged Asnis to send blood and spinal fluid to the state health department lab in Albany for testing.

On Friday afternoon at 4:30, the two women talked again. Asnis's first two patients were failing, and now two others worried her as well. An 80-year-old man had suffered a heart attack and been admitted with a high fever; he too had become too weak to move and was on a respirator. An 87-year-old woman who had come in with a headache, fever, and diarrhea became unexpectedly confused, then paralyzed all over. As Layton and Asnis talked, a neurologist at Flushing Hospital happened to walk into Asnis's tiny office. Overhearing the conversation, he mentioned that he was treating a similar problem at a nearby hospital in Queens—another patient with encephalitis and severe muscle weakness.

In all of 1998, New York City had logged nine reports of encephalitis. Now four, maybe five, cases had sprung up within a week in the same neighborhood. Layton cancelled her trip to the Adirondacks. That Friday night, she hunkered down in her office with medical books, trying to find out what kinds of encephalitis caused diffuse muscle weakness. According to the textbooks, none did. At Flushing, Asnis had also been wracking her brain trying to find a cause. AIDS, Lyme disease, polio, TB, syphilis: nothing seemed too outlandish to consider.

On Saturday, Layton and a health department colleague, Annie Fine, drove out to Flushing Hospital. It was a warm, sunny day, like so many that summer. They chatted about what might be causing the strange outbreak. Fine parked the car on a residential side street. As Layton slid out of the passenger seat, she sidestepped a dead crow on the grass near the curb.

Founded in 1884, Flushing Hospital is a true community hospital, with health fairs and free blood pressure screenings and complimentary dental checkups. It's also an international meeting ground, with brochures printed in Spanish, Russian, Korean, and Chinese—a reflection of the rich polyglot culture of Queens, where 167 nationalities speaking 116 languages make it the nation's most diverse county. But the hospital's insolvency showed. Its pale blue lobby looked like it hadn't been redecorated since the 1960s. Its intensive care unit contained only eight beds. In the narrow hallways of the ICU, Layton and Fine examined patients and talked to families, trying to piece together what the victims, three of whom were on respirators, had in common. They asked where the patients lived, worked, shopped, what parks or beaches they visited, what restaurants they frequented. Other than living in the same two-mile-square area, the patients had very little in common. They were strangers to one another. Their medical charts, however, "told an almost identical story," Layton said. "These were relatively healthy adults who had a febrile illness followed by some GI symptoms followed by the onset of altered mental status, confusion, and gradually this muscle weakness." Another clue also kept ringing in Layton's mind. Several of the patients were tanned and toned. That meant they spent time outdoors—which raised the risk that the infection came from mosquitoes.

As Layton and Fine slowly paged through the thick charts and compared notes, they learned that a 57-year-old man with suspicious symptoms had just been wheeled in. Layton walked over to take a look. High fever, hallucinations, fighting so fiercely with the attendants he needed to be restrained: he looked like a classic encephalitis case. Was the outbreak snowballing?

Layton felt queasy. "I call it my gut meter," she said. "It was maybe one-plus acidic on Monday. It was eight-plus acidic on Friday. By the time I left the hospital on Saturday, it was off the scale."

Arbovirus Anonymous

From the seventeenth through the early twentieth centuries, infections ferried about by insects—malaria, yellow fever, plague, and typhus, to name the Big Four—killed more people and sowed more disease than all other causes combined. At the turn of the last century, discoveries about the natural history of these organisms promised a health revolution. But in the last 25 years, the revolution petered out. Today, 500 million people are infected yearly with mosquitoborne diseases alone; nearly three million die.

What those diseases have in common, along with the infection that struck New Yorkers in 1999, is that the agents that cause them swim in the blood of arthropods. The largest phylum in the animal kingdom, arthropods are segmented creatures with jointed legs and a tough exoskeleton made of chitin. Among Arthropoda's uncuddlesome members are lobsters, crabs, centipedes, cockroaches, and scorpions. For humans, the deadliest members of the clan are mosquitoes, ticks, and fleas.

Viruses transmitted by these blood-feeding arthropods are called arboviruses: *ar*thropod*b*orne *viruses*. And though arboviruses usually cause just mild, flulike infections, they can also kill strapping adults in a matter of days, by playing havoc with the central nervous system, the heart, or the liver and kidneys. Unlike viruses that cause chronic infections, such as the herpes virus, or that spread person to person, like measles, arboviruses don't need people to propagate; in the human body, they are accidental tourists. And because these viruses haven't evolved to the point where they can call a truce with human biology, they often end up knocking off their hosts.

However short-lived, the vanquishing of the great arthropodborne diseases was a monumental achievement in public health. For millennia, sharp observers had noted that biting insects and disease go together. By 1848, scientists surmised that such devastating killers as malaria and yellow fever were transmitted, not by personal contact or fetid water (the usual suspects), but by mosquitoes. When the Scottish physician Sir Patrick Manson showed, in 1877, that the female *Culex* picked up in its bloodmeal the larval form of the roundworm causing elephantiasis, it opened a rich new scientific lode. But not until the discoveries of Ronald Ross and Walter Reed—two of the 29 names chiseled on the frieze of the New York City Department of Health building—was the link made. Their work dramatically changed the course of public health and of history itself.

In 1898, the British protozoologist Sir Ronald Ross demonstrated in India that the malarial parasite *Plasmodium* whiled away part of an exquisitely complicated life cycle in the stomach of the "dapple-winged" *Anopheles* mosquito before entering the human bloodstream by bite. For this insight, Ross won the 1902 Nobel Prize in medicine. (He did not live to savor the award; the year he became a Nobel laureate, the 51-year-old Ross died of appendicitis.) Trench building for the Panama Canal, begun in 1879, had ground to a halt because of malaria and yellow fever. Thanks to mosquito control measures inspired by Ross's discovery, and to quinine, the channel was completed in 1914. More important, millions of people living in tropical climates escaped the parasitic disease.

Around the same time, U.S. Army Major Walter Reed tackled the equally frightening scourge of yellow fever. That virus had come to the Western Hemisphere in the sailing ships of slave traders, as the highly adaptable mosquito *Aedes aegypti* found water cisterns a perfect place to lay eggs. For centuries, the best medical minds didn't make the connection. After Philadelphia's devastating 1793 outbreak of yellow fever, the eminent colonial physician Benjamin Rush concluded that the epidemic sprang from putrid coffee unloaded on the docks. Not until the turn of the last century did Reed's Yellow Fever Commission set up shop in Cuba and reveal the mosquito connection. To test the prevailing theory—which held that yellow fever spread on contaminated articles of clothing or bedding, known as "fomites"—Reed's team built a room in which volunteers slept for three weeks on cots and in bedclothes soiled with waste and body fluid spewed from every orifice by yellow fever victims. The unimaginably stoic recruits emerged in the pink of health. The human guinea pigs who stayed in a virtually sterilized room—where all domestic items had been disinfected by steam but where mosquitoes that had recently fed on yellow fever victims were let loose—sickened within days.

By 1910, scientists had worked out the life cycles of the bloodsucking carriers of dengue fever, typhus, Rocky Mountain spotted fever, African sleeping sickness, and other diseases. Still, in the 1930s, as many as 600,000 Americans got sick from the locally transmitted malaria. During World War II, the U.S. Public Health Service established in Atlanta, the heart of the American malaria zone, a unit called

Malaria Control in War Areas—largely to protect troops in training and assure the continued production of war matériel. Teams of doctors, entomologists, and engineers toiled together to tackle the problem, a mission soon expanded to include control of *Aedes aegypti*, the mosquito species that transmits yellow fever. From this wartime project grew the CDC.

Mosquitoes transmit viruses as part of an intricately choreographed dance in nature. To borrow a locution favored by post-structuralist academics, a mosquito's success as a disease-bearer is highly "contingent." It begins when a female mosquito bites a warm-blooded animal, usually a bird but sometimes a rodent or other animal. (And yes, it is always the female, who needs proteins in the animal's blood for her eggs; male mosquitoes are content to dine on nectar.) Should the bitten animal—known as a "reservoir host" for the virus—be infected, the virus will enter the female mosquito's gut, where it reproduces. The virus then travels to the mosquito's other organs, including the salivary gland, where it multiplies some more. Within a few days of her original bloodmeal, the mosquito infects the next victim—human or otherwise—on which she feasts. She is known as the "vector," or carrier, of the disease-causing virus. Humans have earned the sobriquet "dead-end hosts," because they generally don't carry enough virus in the blood to infect the next mosquito that bites them, which means the chain of transmission is cut.

But mosquitoes are not merely flitting hypodermic needles. They don't promiscuously poke their proboscis into any inviting piece of flesh. Many, in fact, don't bite people at all. They prefer particular species—those reservoir hosts, such as rodents or birds or other animals—that don't get sick. For a few days, reservoir hosts develop high enough quantities of virus in the bloodstream (a condition known as viremia) to pass on the virus to any other mosquito that happens to bite. Mosquitoes themselves remain infectious for life. In a kind of viral chain letter, mosquitoes bite viremic reservoir hosts, which in turn infect more mosquitoes, which bite more animals. Transmission spirals higher and higher. The more a mosquito bites people rather than animals, of course—following with her antennae invisible plumes of exhaled carbon dioxide, body heat, and perhaps a special bouquet of lactic acid and other chemicals from a human target up to 100 yards away—the better the chance of spreading disease. Arboviruses' most extraordinary trick is their ability to breed in both cold- and warm-blooded species, the upshot of viruses and their mosquito hosts evolving in tandem. So ingenious is arbovirus evolution, in fact, some viruses actually disable the mosquito's salivary glands, making it harder for the insect to draw blood, so that it must probe many more hosts, spreading the virus far and wide.

All of which is to say there is an irreducible complexity and an inherent mystery in arbovirus disease outbreaks. So tangled are the factors that permit arboviral diseases to thrive—including temperature, rainfall, population densities of mosquitoes and victims, mosquitoes' or animals' susceptibility to infection, and human behavior—scientists haven't been able to devise good forecasting models for outbreaks. "There have been a lot of very smart people working to predict risk of mosquitoborne disease for a long time," says Roger Nasci, a research entomologist with the CDC. "Nobody's even come up with the easy stuff."

House to House

On Sunday morning, August 29, Marci Layton put through a call to John Roehrig, chief of the arbovirus diseases branch at the CDC's outpost in Ft. Collins, Colorado. Ft. Collins is the World Health Organization's arbovirus reference center for North and South America—a specialized lab to which other labs can send specimens to help them identify a virus or confirm a diagnosis. In his 20 years at the CDC, Roehrig had never received a weekend emergency call. Roehrig, after all, was a lab man. On this Sunday, he was preparing to watch a Denver Broncos preseason game.

Layton asked if the strange cases at Flushing Hospital could be caused by an arbovirus. Roehrig didn't think so. Arboviruses don't usually cause muscle weakness. Besides, there hadn't been any outbreaks elsewhere in the Northeast.

By that Sunday afternoon, Layton and Fine had flushed out three more cases, all from the same Whitestone, Queens, neighborhood, bringing the official count to eight. Having already ruled out common summer viruses, Layton was starting to focus on St. Louis encephalitis virus: a mosquitoborne arbovirus never before diagnosed in New York City. First identified in the 1930s, SLE has since hopscotched erratically around the country, occasionally erupting in large, deadly outbreaks. Normally, the CDC gets 30 or so reports a year. In 1975, SLE shocked public health officials when it sickened more than 1,200 victims and killed 104 in the United States, mostly in the middle of the country.

If an arbovirus epidemic had alighted in New York City, it would be the first since the yellow fever days, two centuries earlier. But it wasn't the epochal quality of mosquitoborne disease in New York City that would make a public health official quail—it was the practicalities. For starters, 7.4 million residents were susceptible. Mosquito spraying would have to start immediately. But New Yorkers, unlike residents in southern U.S. cities, never had to face the seasonal ritual of mosquito control—the spraying, the fogging, the daily squirts of mosquito repellent. Living alongside nature required a certain humility, a trait not prominent in the New York psyche. To top it off, New York City had years before abandoned its mosquito surveillance program. To scout out the vectors of SLE, the city health department had to borrow an entomologist from the venerable American Museum of Natural History.

On Wednesday, September 1, a woman named Varuni Kulasekera toured Whitestone, Queens, to search for mosquito breeding sites. Kulasekera specializes in the *Culex pipiens* mosquito, a drab brown bloodsucker commonly known as the northern house mosquito. In November 1998—nine months before Flushing Hospital's mysteriously paralyzed patients—Kulasekera had received a call from the city on a different but equally curious matter. As the holiday season approached, residents on Manhattan's old-money upper east side, not far from the mayor's residence, Gracie Mansion, were complaining about mosquitoes. Though snowflakes flew outside, people were running air conditioners to freeze out the buzzing menace, hanging gauzy fabric over their beds, and rubbing citronella on their skin at night. On the day before Christmas, Kulasekera stood in a snowstorm incongruously clutching a mosquito net. With a TV crew recording the action, city workers opened a sewer cover. Hundreds of mosquitoes swarmed out. Even in the whiteout, she recognized them instantly as *Culex pipiens*. "If they are still alive at this time of year," she recalls telling a TV reporter, "then we have to be very careful next summer. We might have a new arbovirus outbreak."

Kulasekera, 42, was something of an entomology prodigy. Having grown up in a well-to-do family in Colombo, Sri Lanka, she had rejected her parents' expectations that she would become a doctor and instead followed her passion for insects—a passion kindled at age seven when she found an Atlas moth laying eggs on one of her storybooks. While her friends kept dogs and cats and parrots as pets, Kulasekera nurtured giant water beetles and water bugs in an aquarium, studying their life cycles. At a high school science fair, she displayed beetles, butterflies, and moths collected from her backyard. Her exhibit, which earned the school an award, was titled "Insects from My Garden." After college, she did graduate work on rainforest mosquitoes, living in an open hut in the jungle. "I remember my uncle telling my mom, 'You can never arrange a marriage for her.' I was always an outcast. I used to dream about going to America."

In 1986, Kulasekera did just that, joining the Smithsonian Institution. At the University of Maryland, she dove into what she laughingly calls the most "macho" field in entomology: systematics, which embraces evolutionary biology and taxonomy. The joke is that Kulasekera is a frankly, proudly glamorous woman, an effect abetted by high spiritedness. With large dark eyes and thick curly hair, a gold post in her right nostril, she carries off body-hugging short skirts and high-heeled sandals and dazzling accessories both in the lab and in the field. To cap the larger-than-life, only-in-New York effect, and the gender joke, her husband happens to be the bassist for the quirky cult band Violent Femmes.

On September 1, Kulasekera inspected the Whitestone homes of the known encephalitis victims, looking for clues. After a mild winter and rainy spring, the summer of '99 had been hot and dry. *Culex pipiens,* as it happens, likes drought conditions, which leave stagnant water with lots of organic debris. Septic tanks, rain gutters, and storm drains filled with rotting leaves offered plentiful nutrition to *Culex* larvae. In one backyard Kulasekera saw thick grass and bushes, an ideal site for *Culex.* Another home kept a birdbath, in which Kulasekera found mosquito larvae. At a third house, Kulasekera was puzzled to find no traces of mosquitoes—but in broken English, the patient's wife explained that she and her husband often walked along the East River. When Kulasekera's team rooted around where the woman had pointed, they found dozens of discarded tires hidden under straw, above which *Culex* laced the air.

Late that morning, Kulasekera knocked on the door of a fourth home, and an older woman came to the door. She and her family were about to drive to Flushing Hospital. There she would say a final goodbye to her 80-year-old husband—the man who had shared her life for more than 50 years—before asking doctors to remove a mechanical respirator. All their life together, he had been so glowingly healthy and robust that his wife had always assumed she would die before him. She escorted Kulasekera to the backyard. On one side was a brilliant flower garden, which the wife had tended. On the other was a lush patch of vegetables, which her husband had cultivated. Scattered around the yard were five-gallon white plastic buckets for collecting rainwater, her husband's Good Samaritan gesture during the summer drought. Recently, the wife said, her husband had been suffering insomnia. To keep from waking her, he would sit outside in the middle of the night and smoke. Kulasekera dipped a ladle in a plastic bucket near the man's chair. Floating on the surface were scores of *Culex* larval cases. In her report to the health department, Kulasekera described Whitestone, Queens, as the perfect ground zero for an SLE outbreak.

A Diagnosis

On Wednesday, September 1, the 80-year-old Good Samaritan passed away. The next day it was an 87-year-old woman, below whose second-floor apartment a nephew had also collected rainwater. By then, the New York state health department laboratory had found that blood and spinal fluid from the initial cases tested positive for St. Louis encephalitis virus. Because of the potential scope and public health implications of the outbreak, CDC–Ft. Collins would have to confirm those astonishing results before the city health department could announce the news.

To Marci Layton, the SLE diagnosis was a relief. In public health, infinitely worse than rushing into action against a known enemy is holding fire until one learns the enemy's name. "It's the not knowing," said Layton, that keeps her up at night.

If it did turn out to be SLE, the city would have to set up a full-court press to stem the outbreak and reassure the public. With so many cases piling up—there were now more than 30—and a good percentage of those cases either dead or in dreadful shape, the virus looked to be extremely virulent. In a matter of hours, city officials would have to buy directly from the major pharmaceutical supply companies what would amount to nearly the entire national supply of insect repellent— a good half-million cans and bottles. They would have to prepare airplanes and helicopters and trucks to blanket the city with insecticide— while simultaneously securing permission from the state's Department of Environmental Conservation to spray. They would have to build a website and set up a 24-hour public hotline and write educational flyers in eight languages. They would have to recruit hundreds of city workers to walk through neighborhoods to distribute those flyers. They would even have to make sure the mayor's van was parked at ground zero in Queens, so that he could stage a press conference to announce the city's response.

At 1:30 p.m. on Friday, September 3—the day before the Labor Day weekend—Layton got the CDC's call: positive on SLE. Annie Fine ran through the hallways on the department's third floor, shouting like Paul Revere, "It's positive! It's positive!"

At 3 p.m., Mayor Rudolph Giuliani held a press conference, trying to wrap his rhetoric around an unaccustomed topic. (A week later, he was more pithy and assured: "The more dead mosquitoes, the better.") Within hours, helicopters began spraying in northern Queens, close to the Whitestone Bridge, where the first fatalities had lived. After the press conference, which Annie Fine attended, neighborhood residents walked up to tell her they'd seen a lot of dead robins and crows in the area, and asked if it could have anything to do with the human cases. Fine didn't know. That evening, she helped staff the public hotline during its first hours of operation. A woman calling from Queens had also complained about dead birds, and mentioned a *Queens Chronicle* article on the avian epidemic that had just run. Fine asked her to read it over the phone. The story described dead birds not just in Whitestone but in Manhattan, the Bronx, Long Island, and Westchester County. Two weeks earlier, a *Chronicle* article on the die-off had led with a question: "Has a plague hit the Bayside area?"

Fine felt a shiver of fear. If the bird deaths *were* related, it suggested a much bigger human outbreak than the one seemingly confined to a small peninsula in northern Queens. As if to confirm her apprehension, human cases had turned up just hours before in the Bronx. That night, Fine called city and state experts to ask if the bird deaths could be tied to the human cases. They repeated the conventional wisdom—SLE doesn't kill birds—and she tried to put the matter out of her mind.

What Fine didn't know was that crows had been dropping for months. During the previous decade, the city's crow population had steadily mounted as warmer winters invited the scavenging birds to stick around. Now, in every borough, these raucous urban denizens were suddenly lifeless and littering the sidewalks. Since June, a veterinarian at a clinic in Bayside, Queens, had been treating eerily docile corvids. One woman brought in a crow perched on her shoulder like a parrot. At Ft. Totten, a Civil War battlement in Queens, guards found crows stumbling around as if drunk. That August, highway crews had hauled into the state's Department of Environmental Conservation plastic bags filled with the glossy black bodies of more than 400 dead birds.

Marci Layton and Annie Fine live near each other in a brownstone area of Brooklyn. That night around 2 a.m., after nearly 24 hours of nonstop frenzy, they sat on the stoop of Fine's apartment building. "I remember us hugging and almost crying," Layton said, "because we had been there from the beginning together. I said, 'I know it sounds crazy, but I am so relieved.' Because I knew what it was. I knew what to do."

Fine also felt exhilarated. This is what a public health professional dreams of: to catch the wave of a big outbreak. But even in her exhilaration, something didn't feel right. The dead birds. *Had* a plague hit New York City?

Second Epidemic

On August 9, 1999, three days before Deborah Asnis had seen her first puzzling encephalitis patient, Tracey McNamara, head of the pathology department at the Bronx Zoo, had begun noticing dead crows on the zoo's grounds. As McNamara knew, it takes a lot to kill a crow. Connoisseurs of everything from restaurant scraps to road kill, they are able to fend off most pathogens in nature. So hardy is this species, wildlife disease archives had never recorded a crow die-off in the United States. McNamara sent some of the crows to the state's Department of Environmental Conservation. For two weeks she phoned, eager to hear the results of laboratory tests on the birds. As it turned out, no one had performed lab tests. The state's pathologist simply cut open the bodies and examined them superficially. He told McNamara he found signs of metabolic bone disease—certainly nothing that could explain a die-off.

As the weeks passed, more dead crows turned up on the zoo grounds. Then some of the zoo's pheasants started to show neurological oddities. Worried, McNamara launched her own laboratory investigation. She cut sections of brain, heart, and other organs. She made tissue slides, performed bacterial cultures and stains and toxicology tests. She froze tissue, embedded it in paraffin blocks, submerged it in formalin.

She went through these paces because, in an age in thrall to molecular biology, the distilling of disease down to the level of genetic switches, McNamara resolutely practices old-fashioned pathology with a Zenlike concentration (away from the office, she's a glass-blower and Buddhist meditator). "I cannot make any assumptions. I look at every tissue every time. Every animal that dies in our facility receives a full postmortem." Such is the culture of veterinary pathology, a throwback to the earlier era of the generalist, not the specialist. "Expecting the unexpected is a routine part of our job," she said. "To be a good pathologist, you need a fundamental level of humility, to be able to challenge your own diagnosis. You have to have a sense of inner stillness. The mental discipline to observe and not jump to conclusions." When her initial examination uncovered subtle brain lesions, she concluded she was looking at a viral encephalitis and continued the postmortems.

On September 3, McNamara heard news reports about St. Louis encephalitis in the city. That Labor Day weekend, she pored over textbooks and filled several legal pads with notes. The books repeated what she and apparently everyone else already knew: birds don't die of SLE. And flaviviruses—the group that includes SLE—had never been a veterinary problem. Yet she was sure that what was happening in humans was linked with what she was seeing on her laboratory slides. "Encephalitic people, encephalitic birds," she said. "Could there be a connection? I didn't have to think about it too long."

And luckily, she didn't. Her hunch led her to the most crucial evidence in the outbreak, the key that could have unlocked the mystery weeks earlier than it was solved, had she not been dismissed as, in her own stung interpretation, "some dingbat, premenopausal female veterinarian in New York City."

Large and energetic, her blond hair combed back in a practical style, sharp green eyes registering everything from behind dark-rimmed glasses, McNamara has natural comic timing. Described by her peers both as "brilliant" and as "entertaining"—not a common adjectival pairing in the scientific world—she tends to shoot from the lip, which both catalyzed and scandalized the new community of scientists that would soon surround her.

McNamara's earliest childhood memory is of visiting the Central Park Zoo with her parents. In those days, the 1950s, the director of the zoo reared a Kiplingesque black panther and even walked the cat on a leash through the park. When McNamara walked up to his cage, the panther was pacing back and forth. "I said 'Bagheera?' And the cat just stopped in his tracks and came up to the front of the cage and looked at me. That was it." At age three, she had fixed on her life's work. During college, after a detour into French literature, McNamara took off for Kenya to labor free for a team of wildlife researchers. At first, they turned her down, for lack of experience. "I sat outside of people's doors for eight hours a day. I wore 'em down. I just wouldn't take no for an answer." Eventually, she found herself rattling across the savannah on the back of a pickup truck, helping out on a study of giraffe physiology.

By 1999, at the age of 45, McNamara had risen to the top of an elite professional cadre. She is a veterinary pathologist specializing in wildlife diseases, one of perhaps ten such doctors of veterinary medicine in the country. "Veterinarians enjoy this reputation of James Herriott," she said. "You know: all creatures great and small, veterinarians take care of them all. Well, baloney. The reality is more like George Orwell's *Animal Farm*: all animals are created equal, but some animals are more equal than others. For centuries, the emphasis has been on domestic species, on the economically important species. As a result, even today in the veterinary curriculum, more time is spent on the dog and the cat than on the whale or the bat." Diagnosing diseases of wildlife—animals that are literally no one's business—places one on the scientific frontier. Or perhaps outback. Much of what wildlife pathologists see, they are seeing for the first time. At her office in the Bronx Zoo (now formally known as the Wildlife Conservation Society), McNamara receives samples from field researchers all over the world. She discovered a herpes virus in Malayan peacock pheasants, a cerebral fungus called *Cryptococcus* in the common anaconda, and a pox virus in South American ruminants known as pudu.

On September 7, the day after Labor Day, McNamara arrived in her office to learn that a Chilean flamingo and a Guanay cormorant had died and more zoo birds appeared sick that morning. The zookeepers were upset—they knew these animals like friends. McNamara performed immediate postmortems. In the flamingo, she found massive myocarditis ("heart transplant city"), in the cormorant extensive hemorrhaging in the cerebellum. "The most frightening lesions I have seen in sixteen years as a comparative pathologist," McNamara says. "Hair stood up on the back of my neck . . . I could feel in my bones it wasn't anything that any of us knew." Whatever was killing the birds, it looked "hot"—lethally infective. Donning a respirator mask and goggles, a disposable jumpsuit and three sets of gloves, she proceeded to do her handiwork under a filtered-air hood. With a new razor blade, she cut fine sections of the fresh, Jell-O-like organs. She took impression smears—a kind of outer cell scraping of the organs, from which a pathologist can quickly characterize cell damage. She took three sets of tissue samples from every organ. She prepared 40 tissue slides for her technician. The next day, the technician called her out from the necropsy suite. Peering at wet sections of tissue sliced 6 microns thick, she saw "the worst meningoencephalitis I have ever seen. And meanwhile, I have more dead birds in the cooler. I have a pheasant with massive cardiac necrosis and hemorrhage. I have a snowy owl with acute coagulative necrosis in the liver. I have two flamingoes on September 9 with GI lesions." McNamara was desperate. The zoo had never seen anything like it: birds from distant evolutionary lineages apparently struck down by the same disease. Over the next few weeks, 24 zoo birds would die or be euthanized. The victims would include black-billed magpies, bronze-winged ducks, a northern bald eagle, and a black-crowned night heron. Most had unmistakable central nervous system damage. One flamingo craned its neck low in a weird S-shape. Others staggered on wobbly legs. Laughing gulls drooped their heads. A cormorant swam deliriously in circles.

McNamara mulled the problem like a detective sifting clues, drawing on both logic and intuition. Her thinking went something like this: It could be avian influenza or Newcastle's disease, but those didn't quite fit the lesions she was seeing. Besides, those viruses wipe out chickens and turkeys—but in the children's zoo, the farm birds were doing just fine. Maybe it was an alphavirus, of which the most likely would be Eastern equine encephalitis. EEE, which can be fatal to humans, is mosquitoborne, and McNamara knew the zoo was having a mosquito problem because there had been cases of avian malaria. But EEE is notorious for killing emus, an Old World species, and the zoo's emus were thriving. What most bothered McNamara was that the only birds the zoo was losing were North and South American species. Normally, those would be the very species that would survive and act as reservoirs for the known arboviruses of the Western hemisphere, such as St. Louis encephalitis.

On September 9, McNamara phoned the CDC's offices in Ft. Collins. She had a pressing concern on her mind: one of her veterinarians had stuck himself with a needle while euthanizing a flamingo. Surmising that the animal and human outbreaks were related, she wondered: What kind of strange infection might he have caught? But she also wanted the CDC to know that the resonance between SLE in people and what was killing her avian charges was striking, and that exploring one might yield clues to the other. McNamara asked CDC to test serum from the veterinarian who had stuck himself and plasma from the dead flamingo. A biosafety officer in Ft. Collins flatly turned down her request, in part because birds weren't the CDC's business or subject of expertise. As McNamara remembers, "He cut me off in mid-conversation and said, 'Birds don't *die* of St. Louis encephalitis. Birds are a *reservoir* for St. Louis encephalitis. And you should know better. You're just dealing with some veterinary thing.'" McNamara's jaw dropped. She sent the serum anyway.

That day, McNamara also called the National Veterinary Services Laboratory, in Ames, Iowa, part of the Department of Agriculture. Since their purview was animals, would they test her specimens for encephalitis viruses? To her relief, NVSL said yes. That weekend, their lab workers called McNamara at home. The zoo's specimens were negative for Eastern equine and other alpha viruses. But something was growing in their cultures—something they couldn't identify. A few days later, when the NVSL called again, McNamara was in the zoo's necropsy suite, wrist deep in more postmortems. She pulled off a glove and grabbed the phone. The mysterious virus had finally started growing. Under an electron microscope, it looked to be about 40 nanometers in diameter: too small for an alphavirus like Eastern equine encephalitis, but just right for a flavivirus such as St. Louis or others. The NVSL didn't have the technology to further pinpoint the virus.

McNamara felt flushed with triumph. Her intuition was right: there *was* a connection between the bird and human cases. "That validation of diagnostic skills. Everything I'd ever studied, everything I've ever been trained to do—it paid off. It was right." But after that surge of elation came a jolt of fear. Maybe it was SLE or maybe it was a more dangerous viral cousin. Suddenly it dawned on her that she was slicing up tissue loaded with an unknown flavivirus, and that she was relatively unprotected.

McNamara quickly phoned Ft. Collins to tell them about the NVSL's discovery of an unknown flavivirus in the zoo's birds. By now, three of her lab workers had stuck themselves with needles, and she was frantic to know exactly what might have infected them. Birds were piling up in her cooler. Her lab looked like a field hospital in the midst of what public health officials had declared a war zone. Yet despite her best efforts, she couldn't talk to the generals.

Lab Work

In the fall of 1999, the list of identified arboviruses numbered 538. Of this list, those known to cause human disease ran to 110, and most had niches far removed from the United States. And for these 110, diagnostic tests existed for fewer than 30.

When the New York City epidemic broke out, lab workers at CDC–Ft. Collins did the practical thing: they tested for the viruses most likely to strike in that geographic area. In early September 1999, the protocol was to initially look for four arboviruses: LaCrosse, Powassan, Eastern equine encephalitis, and St. Louis encephalitis. And that's just what Robert Lanciotti did. Lanciotti, a tall, broad-shouldered man with reddish hair and a faintly military air, directs the branch's diagnostic and reference labs. He is respected as a calm, methodical professional, the kind of person you'd want around during an outbreak.

When he received serum and tissue from the first group of patients diagnosed in New York City, Lanciotti ran a blood test known as an IgM and IgG ELISA. IgM is the antibody the immune system produces when a person is acutely ill. According to the textbooks, IgM appears a few days after an illness begins and sticks around for 30 to 45 days or more, then vanishes. By contrast, IgG antibody persists for a lifetime. If somebody is IgM negative and IgG positive for a particular virus, it means the person was infected at some indefinite point in the past. When Lanciotti tested the New York City samples against a battery of four Eastern viruses, the sample came back both IgM and IgG positive for St. Louis encephalitis, the most widely reported mosquito-borne disease in the United States. He considered it a "presumptive case of St. Louis."

It was "presumptive" because SLE is a flavivirus, and since the 1950s scientists have known about a peculiar hitch in antibody testing for flaviviruses. Far-flung and powerful, like a viral Mafia, the flavivirus family includes the agents that cause yellow fever and dengue fever. (A distant flavivirus cousin, the deadly hepatitis C virus, never evolved in arthropods.) Like a dutiful clan, flaviviruses sport nearly identical proteins on their surfaces. As a result, they cross-react to antibodies targeted for other family members—meaning that serum containing one flavivirus could easily come up positive in tests for other flavivirus kin.

Years ago, virologist Karl Johnson described the problem as a "hall of mirrors." When Rob Lanciotti got a positive result for SLE, all he knew for sure was that the New York City patients were infected with some sort of flavivirus. The typical confirmation is known as a neutralization test, which reveals whether antibodies in a patient's serum actually attack the live form of the suspected virus. The CDC began its neutralization tests, which take a week or two for flaviviruses, at the beginning of September.

Serum antibody tests are good in a pinch. But the gold standard for identifying a human virus is to grow it in other mammalian cells, such as monkey cells or in a chicken egg—a technique known as virus isolation. After that, the virus can be pinpointed with antibody or antigen tests, or with polymerase chain reaction, which amplifies sections of the genome. Unfortunately, the CDC was never able to grow the suspected virus from the clinical samples it had received. Nor had lab workers expected to. By the time symptoms have begun, there aren't a lot of virus particles left in the bloodstream to grow.

Publicly, the CDC and local health agencies were sticking with the SLE diagnosis. But privately, as September wore on, the diagnosis started to look shaky. One problem was Ft. Collins's serologic tests on new patients' blood samples: the IgM assays—which signal acute infection—looked suspiciously weak. At the New York City health department, those results bothered Marci Layton. She kept quizzing John Roehrig about their meaning. How can we be sure, she asked him, that we're really dealing with St. Louis encephalitis? Meanwhile, CDC officials were learning of dead birds not just at the Bronx Zoo but all over New York City. For weeks, the health department's Annie Fine had been asking CDC officials if the human and bird deaths could be related. Derisively, they cut her off. Feeling like a pest, she eventually stopped bringing it up. But CDC officials were bothered enough to call the nation's leading arbovirologists to ask whether SLE had ever been known to kill birds. (Rarely, said the experts.) The pieces of the puzzle were not fitting.

Finally, in the third week of September, the puzzle pieces assorted themselves neatly—into a new picture. First, the CDC learned that the

NVSL had isolated a flavivirus from Bronx Zoo birds; that bolstered Tracey McNamara's suspicion that what was killing her charges might be killing New Yorkers. Early the same week, the Connecticut Agricultural Experiment Station announced that it had isolated a flavivirus from an encephalitic crow and from two pools of mosquitoes; that evidence drew a stronger thread, since presumably the infected mosquitoes were biting both birds and humans. But were those mosquitoes infected with SLE or something else? When Marci Layton and Annie Fine heard the Connecticut news, "Lights started going off in both our brains," Fine recalls. Their persistent doubts about the SLE diagnosis seemed borne out. Later that week, researchers at the U.S. Army Medical Research Institute of Infectious Diseases, or USAMRIID—to whom Tracey McNamara had sent specimens as a last resort—also found a flavivirus.

On Monday September 20, the NVSL sent their Bronx zoo virus isolates to the CDC. To pinpoint the virus's identity, Lanciotti ran two kinds of genetic tests simultaneously. In one, he used assays that broadly react with either all flaviviruses (such as SLE) or all alphaviruses (such as EEE), or with a third group known as bunyaviruses. In the other, he ran genetic tests that specifically ferret out SLE and EEE. The results baffled him. The broad tests ruled out alphavirus and bunyavirus, and ruled in flavivirus. Yet his test for St. Louis encephalitis came back negative. It didn't make sense. Lanciotti wondered if he was dealing with a strange variant of SLE.

Lanciotti then tried a different tack, comparing a patch of the virus's nucleic acid sequence against the sequences of all known flaviviruses. On Wednesday, September 22, he loaded samples of the virus genome into a machine that automatically sequenced the DNA strands. That evening, to take his mind off the problem, he played golf with some colleagues. The next morning he arrived at work at 6:30, anxious for an answer. Around 7 a.m., he submitted strings of genetic code over the Internet to the National Library of Medicine's GenBank, a database of all publicly available genetic sequences. Ten seconds later, the match flashed on his computer screen: West Nile virus.

A bolt of heat shot through Lanciotti's body. His heartbeat sped

up and he became drenched with
sweat. "It was absolute shock and
horror. I took that data and literally ran upstairs to my boss, and
we ran together up to the division director." Duane Gubler
directs the CDC's division of
vectorborne infectious diseases.
An old hand in arbovirology—
fieldwork on dengue and parasitic
diseases in more than 100 countries—he thought he had seen everything. Hearing the news, all the
wisdom he could summon was,
"Holy shit."

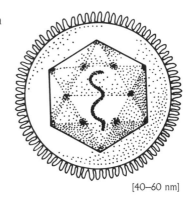

[40–60 nm]

**West Nile virus
(family Flaviviridae)**
Agent of West Nile encephalitis

Global Imports

West Nile virus in New York City. The news was shocking in part
because, in all the thousands and thousands of words poured into scientific papers and government guidelines and prolix draft documents
of all kinds—all those seers scouting out emerging infections on the
horizon—West Nile virus had never come up. That, of course, is the
underlying message in emerging infections: they *are* impossible to predict. West Nile virus, which usually causes only a feverish, flulike illness, isn't the most dire threat from the insect or animal kingdom—
Ebola in Queens easily trumps it. But its breathtaking leap across
continents is a reminder of just how mobile today's pathogens are.
What scares experts is that it's a preview of more dire things to come.

If you were to unfold a map, close your eyes, and point, virtually
any place at the tip of your finger can be reached within 36 hours.
That's less than the incubation period—the time between infection
and the onset of symptoms—for most infectious diseases. In other
words, an infectious agent can slip in and spread before its symptoms

give it away. As cartoonist Gary Larson depicted the situation, "Be a virus, see the world."

Germs are natural vagabonds. Take an epidemic that hit the United States precisely one century before West Nile did. On June 27, 1899, the S.S. *Nippon Maru* had sailed through the Golden Gate into San Francisco harbor. Weeks earlier, after the ship had left Hong Kong, two passengers had come down with bubonic plague. By the time the vessel arrived at its stopping-off point in Honolulu, plague-infected rats were scurrying through the holds. In San Francisco, officials quarantined the vessel, though no passengers had shown any signs of illness. While inspecting the ship, however, officials found 11 Japanese stowaways. Within hours, two of those stowaways were missing. Their sodden corpses, clad in life preservers from the *Nippon Maru*, were later fished from the Bay—full of plague bacilli. Nine months later, San Francisco ushered in the Western Hemisphere's first epidemic of bubonic plague. On March 6, 1900, a body turned up in the basement of a Chinatown hotel. Despite a massive sanitary campaign by fumigation, sponging, and washing, the outbreak lasted four years and killed 112 individuals, mostly Chinese. After the 1906 earthquake, the city's rat population exploded and a second epidemic unrolled. So just a century ago, sailing ships were the distribution channels for bubonic plague's last global pandemic.

Killing 60 to 90 percent of its victims before the antibiotic era, plague had spurred some of the earliest public health laws in Europe, including quarantine of incoming vessels and isolation of victims. Throughout history and up to today, the idea of preventing infectious disease by corralling sick people has had its appeal. In turn-of-the-century San Francisco, the local board of health responded to the plague outbreak with what we would consider impractical, not to mention racist, measures: stretching a rope around the twelve-block Chinatown and stationing police along the line to keep the stigmatized residents inside. Similarly, during the yellow fever epidemic of 1793, bands of vigilantes in New York City patrolled the streets at night to prevent potentially tainted Philadelphia fugitives from slipping in.

But today, neither ropes nor vigilantes, electric fences nor the Marines, can close the door on imported infections. New York City is a pulsing intersection. Between 1987 and 1997, the number of international passengers and cargo flights to and from airports serving the city doubled. In 2000, fully 40 percent of the city's population had been born abroad, the highest percentage of foreign-born residents since 1910, when waves of immigrants surged through the halls of Ellis Island. Across the United States, tens of millions of foreign-born travelers enter each year, along with one million legal immigrants and refugees and several hundred thousand illegal immigrants—many from countries where infectious diseases are rife. Worldwide, more than 1.4 million people travel across international borders by air each day—more than 500 million travelers each year, a figure that grows by 10 percent a year. Though immigration opponents may lean on the cruel stereotype that "foreigners" bring in disease, the fact is that everyone who travels, including affluent U.S. tourists, can bear bacterial, viral, and parasitic gifts.

In a time of unprecedented travel and trade and transport, West Nile virus was just another tourist. And though it captured the headlines during the summer and fall of 1999, another disquieting import had arrived in its shadow. That summer, at Baiting Hollow Boy Scout Camp, a hilly, wooded site bordered by wetland on Long Island, two 11-year-old boys came down with malaria. Not having traveled outside the United States, the boys could only have acquired the disease from the bite of an *Anopheles* mosquito. And the mosquito could only have become infected by biting, about two weeks before, someone as close as half a mile away who had harbored in his or her bloodstream the mature male and female stages, called gametocytes, of the malaria parasite—in this case, *Plasmodium vivax*. Someone experiencing a relapse of the disease years after the initial infection can host malaria parasites in the bloodstream for days or weeks, never experiencing symptoms. In this case, investigators never found that human source, almost certainly a visitor or immigrant from a malarious country.

Nor did public health officials ever find the source of a 1993 malaria outbreak in Queens, that sickened two and perhaps three people who lived within a mile of each other. In those cases, the agent was *Plasmodium falciparum*—the deadly strain of the infection that kills 3,000 people worldwide every day. Many immigrants in the Queens neighborhood had arrived recently from malaria-endemic areas, including parts of South and Central America. What bewildered officials was that crowded urban areas are usually uncongenial for *Anopheles*, which prefers clean, slow-moving water. Each year, one or two such outbreaks pop up in the United States, often where the victims are homeless or living in substandard housing, and thus easier targets for mosquitoes. Just as well-documented is the phenomenon of malaria-infected mosquitoes riding in the cargo holds and wheel wells of jets traveling from tropical to temperate countries, and on their escape biting the first hapless victims they can find—inspiring the terms "airport malaria" or "baggage malaria." (France, host of stowaway mosquitoes from West Africa, is the world's airport malaria leader.)

As a *New York Times* writer put it, "If the indestructible cockroach is the Sherman tank of the insect kingdom, the fragile mosquito is its DC-7, a pioneer in linking the world by air." West Nile virus certainly illustrates that. First discovered in the blood of a fever-ridden woman in the West Nile district of Uganda in 1937, West Nile virus had been tracked in Africa, the Middle East, and western Asia (where, because the disease is endemic, most people become immune in childhood). The branch of the flavivirus family to which West Nile belongs—known as the Japanese encephalitis virus antigenic complex—includes the Japanese encephalitis virus in Asia, Murray Valley encephalitis virus and Kunjin virus in Australia, and SLE in the Americas—distinct geographic niches that have recently expanded and blurred.

But because vectorborne pathogens have become jet age gypsies, what's more surprising than West Nile's arrival is that other disease-causing agents didn't show up here first. Scientists have long been on alert for other pathogenic rovers—and they still can't quite believe a relative unknown like West Nile, which wasn't even on the racing card, crossed the finish line first. Take the virus behind Japanese encephalitis, or JE, which causes symptoms ranging from fever and malaise to convulsions and coma, killing 25 percent of its victims. Once confined to Asia, it has leapfrogged to New Guinea, Australia, and elsewhere in the Western Pacific. During World War II, U.S. officials anxiously monitored the movement of men and matériel from Japan to the West Coast, fearful that the virus would sneak in—or that the Japanese would infect mosquitoes, stow them on one of their submarines, and surface near the United States just to let them loose. During today's peacetime trade, mosquitoes or mosquito eggs from Asia could easily carry in JE, where the virus would then become amplified in birds, as had West Nile virus. In 1998, scientists cringed when they found the Japanese mosquito *Aedes japonicus* in the United States (though, like the notoriously spreading Asian tiger mosquito, *Aedes albopictus*, it probably had arrived earlier, in mosquito eggs deposited in old tires sent here for retreading). "It's just dumb luck that those things were not infected with Japanese encephalitis virus," says University of California–Berkeley arbovirologist William Reeves Sr., "because it can be transmitted from a female mosquito to its progeny." If JE did arrive in the U.S., as many as 30 native mosquitoes could tote it around—and with that many different vectors, eradication would be impossible.

Or consider Rift Valley fever virus, which causes fever, shock, and encephalitis. It could hitch a ride on a viremic person or an infected mosquito from Africa or the Middle East, and would almost surely go on to cause human deaths as well as epidemics among cattle and sheep. In 1977, when the virus jumped from southern and eastern Africa to Egypt, it was said to bear an uncanny resemblance to the biblical description of one of the plagues of ancient Egypt, with more than 200,000 cases and 600 deaths. In 2000, to the dismay of scientists, Rift appeared for the first time outside Africa: in Saudi Arabia and Yemen, now threatening the Arabian peninsula. Could this virus or some similarly undesirable relative come to the United States? "I can't tell you when, but it will," says legendary virus hunter C. J. Peters, now at the University of Texas Medical Branch at Galveston. Citing the propensity of animal pathogens to seek new niches, he says, "If we can have

BSE and foot and mouth disease in Britain, and we can have Nipah in Malaysia, and we can have Rift Valley fever in Egypt and in Arabia, what's different about the U.S.?" An intervening ocean, as well as strong animal quarantine laws, have kept this country safe so far. "But I don't think we can count on blind luck always."

Ross River virus proved in the 1980s that it could escape its niche in Australia to erupt in Fiji and Samoa (probably in a viremic person), where it triggered massive epidemics of fever and arthritis. "Just by luck," says virologist Tom Monath, "nobody got on an airplane" to the United States. If someone had, salt marsh mosquitoes on the coast could spread Ross River virus, as could *Aedes aegypti* and *Aedes albopictus,* now buzzing over half of the U.S. If Murray Valley encephalitis virus jumped from Australia to, say, California, *Culex* mosquitoes, like taxis lined up at LAX, would be ready to transport it.

Not all itinerant arboviruses come to us via mosquitoes from animals. Some are transmitted by mosquito from person to person, with no animal host in the middle, a simpler cycle that makes them potentially more threatening. Dengue fever, the excruciating pain of which has earned it the epithets "breakbone fever" and "devil's crunch," is caused by four closely related tropical flaviviruses; at their worst or when different serotypes cause serial infections, they can trigger massive hemorrhaging or circulatory collapse, especially in young children. The rise of dengue is a modern public health morality tale. Starting in the 1940s, the Pan American Health Organization was determined to eradicate its host, *Aedes aegypti,* from Central and South America—a campaign mainly driven by worries about yellow fever. Unfortunately, the United States never signed on to the effort—several cost–benefit analyses concluded that in a disposable society like ours, the price of cleaning up tire dumps and junkyards was too high compared to the public health gain at home. As a result, whenever a country was successful in eradicating *aegypti*, the United States ended up reintroducing it. *Aedes aegypti* eventually reclaimed North and South America, spreading dengue further than ever and setting the stage for several strikes across the Mexican border into Texas.

But what most worries U.S. scientists is the possible importation of yellow fever, whose mosquito vector is also the thriving *A. aegypti.* Today, the disease strikes about 15,000 victims each year in the tropical regions of South America and Africa. In South America, yellow fever is relegated to the jungle, in a cycle involving mosquitoes and monkeys. Public health officials fear that an infected person—say, a farmer cultivating coca in the jungles of Colombia—could act as a bridge between jungle and city. In a big-city hospital, the patient could infect *Aedes aegypti,* and that could springboard the virus around the world. "It's not a matter of *whether*, it's a matter of *when* we begin to see urban transmission of yellow fever again," says the CDC's Duane Gubler. "I did a little exercise to find out where a person could go from Santa Cruz, Bolivia—where we had some urban transmission a couple of years ago—and it's everywhere. Asia and the Pacific are probably the most important destinations. Even if we don't see major epidemics, I see this as probably one of the next international public health emergencies. You can get on a plane from Santa Cruz and be in Manila or Hong Kong or Bangkok within thirty, thirty-six hours."

Or the United States in even less time. It takes just one infected person to spark an epidemic. In September 1999, as West Nile was seeding itself in American birds and in American paranoia, the U.S. nearly faced just such a crisis. After a 10-day trip to the rainforests of southern Venezuela, a California man—who, unlike his travel companions, had neglected to get a yellow fever vaccination before his trip—returned with fever and chills, blanketed with mosquito bites. Days later, he died a ghastly yellow fever death: seizures, lung failure, and fulminant hepatitis. Luckily, he didn't dispatch the disease to anyone else. But if, during the virus's incubation period, an American tourist in South America had been bitten and returned home—say, to New Orleans—the South could host the first U.S. yellow fever epidemic since 1942. Once that epidemic began to smolder, doctors would probably diagnose it first as infectious hepatitis, because patients' skin would be yellow with jaundice. A pathologist would catch the disease during an autopsy—but since very few deceased patients get autopsies today, several weeks could pass before anyone noticed that yellow fever was in our midst. By that time, squadrons of infected mosquitoes would have spread the infection. Although a vaccine is effective, only about one million doses are on hand in the United States—not nearly enough to put out an urban wildfire—and it would take months before a drug company could gear up production for more. Some scientists say the same advantages of modern living that protect us from dengue fever—window screens, air conditioning, prime time TV—would limit the spread of yellow fever. But others fear a sudden yellow fever outbreak could kill 5 to 10 percent of a city's residents. "It would be about equivalent to what happened in 1918 when we had influenza," says University of Texas Medical Branch arbovirologist Robert Shope, "and everybody knew somebody who died." Or, to put it in a modern tense, it would be about equivalent to what people in poor and neglected countries face every day.

Breaking the News

When CDC officials in Atlanta heard that West Nile had landed on our shores, they grilled Rob Lanciotti about his lab technique: Any chance of contamination? No. Ft. Collins officials phoned Tracey McNamara at the Bronx Zoo. She was in the necropsy suite, as she had been for weeks, cutting up more dead birds. Without revealing that the CDC had identified the virus in her flamingo samples as West Nile, they politely asked her for more bird tissues—FedExed on dry ice, that night, if it was no trouble. McNamara was glad she had finally raised the agency's interest, though she wondered why it had taken so long.

By Friday morning, September 24, investigators in the New York City health department were brimming over with excitement. What exactly had struck their city—a weird new form of SLE? another deadly flavivirus? an enigmatic flavivirus called Rocio, which had killed people and birds in Brazil in the 1970s, then just as mysteriously disappeared? Just when they thought their SLE investigation was winding down, the roller coaster looked like it was about to climb another hill. Dozens of people joined in on a 10 a.m. conference call as CDC notified state and local health officials of the shocking new diagnosis. Duane Gubler opened the discussion by jokingly asking if everyone was sitting down. Gubler informed the group that four birds in New York City—three from the Bronx Zoo and a crow from Scarsdale—had died of what the agency called "West Nile–like virus," to which humans might well be succumbing as well.

Unbeknownst to the CDC, an acquaintance had tipped off Tracey McNamara to the news. Realizing that her bird specimens had unlocked the mystery—samples she had tried in vain for weeks to have tested—she phoned a contact at Ft. Collins, playing dumb, asking how the investigation was going. As it turned out, she had called just in time to log into the conference call. After CDC officials delivered the news, there was a brief pause. McNamara jumped in. "Ladies and gentlemen, my name is Dr. McNamara," she started out. "I'm a pathologist at the Bronx Zoo. And these are my cases we're discussing. Would anyone care to hear the facts?" A meticulous note-taker, she proceeded to read from a log/calendar that she had been keeping since early August: whom she had called, what she was told, where and when she sent specimens. When she was done, she heard a a collective gasp on the line. Annie Fine, who had never heard of McNamara, was dumbstruck. Just a few miles from the health department's headquarters, someone had been holding the key that could have unlocked this public health mystery weeks earlier. For her part, McNamara worried that the Bronx Zoo would be publicly identified as the hot zone epicenter of the outbreak, even though dead crows from all over the state had been turning up as early as July. If so, visitors would stay away from the park in droves. Sure enough, that weekend, after the news hit the press about West Nile–infected birds from the Bronx Zoo, the park lost hundreds of thousands of dollars of business. "You guys play hardball," McNamara told herself. "I'm in the big leagues now."

Later that day, in a fine bit of understatement, Duane Gubler told the *New York Times,* "This is exciting. We prefer it didn't occur, but it is interesting." Luckily, the textbook public health retaliation, which New York City had been practicing for three weeks, was the same for both SLE and West Nile: spraying pesticide to kill adult mosquitoes, eliminating potential mosquito breeding places, applying larvicide to collections of standing water, and advising people to avoid mosquito bites. What public health officials didn't know at the time was that the human outbreak was actually over; the last reported patient had become ill on September 22.

Sneaking In

West Nile virus in America: the finding stirred the blood of nearly every scientist in the country's small corps of arbovirologists and medical entomologists. One was Bill Reeves, now emeritus professor at UC–Berkeley. Back in 1941, Reeves and his team were the first to isolate St. Louis encephalitis from mosquitoes in nature. He was the first scientist to use dry ice to help attract mosquitoes in the wild, now a standard technique. "You know the old story of Dalmatian dogs that live in firehouses?" Reeves said. "When the gong goes off, the Dalmatian runs over and jumps on the firewagon because he doesn't want to miss that car ride. My reaction was very much like the Dalmatian. I wanted to get involved. It sounds silly. I'm 83 years old. I've been retired for 13 years. But when something like this comes, it's like the good old days."

In dog breeding and public health investigations, pedigree counts. To find out where New York's strain of West Nile virus came from, researchers drew a genetic family tree with help from French researcher Vincent Deubel. The strain that hit the United States was nearly identical to one found in 1998 in a farm goose in Israel. Israeli geese probably caught the virus from migrating birds such as storks, whose annual excursions can take them from the northern tip of Europe down to South Africa. (Superimpose a map of stork migration routes over Old World locales where West Nile epidemics have popped up recently, and the overlap is striking.) So while discovered in Israel, the New York strain may have originated anywhere throughout the Mideast and even Africa.

Whatever its provenance, how did the virus beat a path to New York City?

We'll never know. "It's a series of unlikely events—each of which requires a series of unlikely events," says USAMRIID entomologist Mike Turell. The main schools of thought nominate mosquitoes or humans as the transatlantic smuggler.

Mosquitoes, as we've seen, can survive on airplanes and breed on ships. Once in New York, an infected mosquito, like most famished tourists, probably dined at the first establishment that looked inviting—maybe a crow—which in turn infected other mosquitoes, inexorably amplifying the virus. Playing on the notion of "airport malaria," Turell believes the outbreak began as "airport West Nile." And he suspects it wasn't the first time the virus sneaked in. Other infected mosquitoes may have entered from Israel or elsewhere in the previous year, only they couldn't spread the virus. Maybe they were eaten by a dragonfly. Maybe they were caught by a spider. Or maybe an infected mosquito in search of a bloodmeal settled for a dog instead of a bird or person, halting the chain of transmission. All scientists know for sure is that in the summer of 1999, at least one infected mosquito bit a suitable reservoir host. After that, nothing could stop the virus. If the main vector in the United States turns out to be *Culex pipiens*, this viral amplification probably took place under cover of darkness, when crows and other carriers roost communally—*C. pipiens* is a nocturnal feeder. Perhaps the virus multiplied in the Jamaica Bay Wildlife Refuge, adjacent to Kennedy International Airport, a prime stop for migrating birds. More likely the epicenter was northeast Queens, where in 1999 a chilling 50 percent of wild birds became infected. As transmission accelerated, the virus crept closer and closer to people. "It's a numbers game," says CDC vertebrate ecologist Nick Komar. "How many mosquitoes are out there? How many are infected? What's the probability that they feed on a human being? How many human beings are there? If it's really hot, are people sleeping outdoors or spending longer hours outdoors? All those numbers come together. With a huge population like New York's, you're bound to get human infection."

On the other hand, maybe the original vector wasn't a mosquito at all. Maybe a human unwittingly conveyed the virus to the New World. Between July 1998 and June 1999, nearly five million people arrived from international destinations to one of New York City's three airports (La Guardia, JFK, and Newark International). Of those passengers, nearly a quarter came from one of the 59 nations where West Nile virus had appeared in humans or animals. If a newly infected traveler happened to land in New York during the virus's three- to five-day incubation period, during which time virus in the blood peaks, he or she could have infected a big-city mosquito and kicked off the cycle. Even healthy people symptomlessly infected with West Nile, many virologists say, can raise enough viremia to infect mosquitoes. "My suspicion," says Tom Monath, the CDC's former division director in Ft. Collins, now vice president of the biopharmaceutical company Acambis Inc., "is that this was somebody who traveled from Tel Aviv or somewhere to visit a cousin in Queens, sat on a stoop in the late afternoon, and got bitten by *Culex*."

Edging toward the outer bounds of probability, imported or smuggled birds may have spirited in the virus. In 1999, 2,770 birds, both commercially bred stock and pets, were legally imported through JFK. Most imported birds are quarantined for 30 days—unless American owners had taken the birds abroad while traveling, then brought them back. Here, these feathered family members get away with a kind of house arrest, monitored at home by their owners. Maybe a bird went to Israel and on its return infected a mosquito in Queens. CDC investigators went so far as to track down several New York City parrots that had made just such a sojourn. Less likely, but not impossible, is an infected migratory bird having flown a wayward path across the Atlantic from Europe. Even ticks, which are hardy travelers, can carry West Nile virus. The closer you look, the more almost anything seems plausible—and the more tightly the globe seems strung together by viral connections, like one of those busy, loop-decorated route maps in airline magazines.

In October 1999, Richard Preston published an article in *The New Yorker* speculating about whether West Nile had been planted by a bioterrorist. He cited an excerpt from an article that had appeared that April in a London tabloid, ostensibly written by an Iraqi defector claiming to be Saddam Hussein's double. According to the account, the Iraqi president had boasted that his country's labs had perfected a West Nile strain "capable of destroying 97 pc [percent] of all life in an urban environment." The notion of a terrorist depositing an exotic pathogen into New York wasn't new, of course. Since 1995, the year a Japanese cult released nerve gas on the Tokyo subways, New York City officials had been bracing for biological warfare. But it wasn't just intentional malice that they feared. The 1994 outbreak of pneumonic plague in India and the 1995 epidemic of Ebola hemorrhagic fever in Zaire put them on guard for any imported epidemic. When federal investigators learned on September 3 that SLE was afoot in the city—itself a bizarre event—they reflexively began looking into terrorism. When further tests revised the finding to West Nile, more alarms went off. Stoking their suspicions were reports that summer about a West Nile epidemic in southern Russia—a country known to have fanatically perfected bioweapons under the former communist regime.

Ultimately, however, federal officials discounted the bioterrorism theory. West Nile virus, they concluded, not only doesn't wipe out enough people to be a credible weapon, it depends on the tenuous life cycle of mosquitoes to spread: not a reliable way to sow panic. When scientists later mapped the genetic sequence of the virus, they saw that, far from being a devilish recombinant, it could only have descended by natural mutation from the Israeli 1998 strain.

But some researchers aren't willing to completely close the door on the bioterrorism hypothesis. USAMRIID scientists, while skeptical that the West Nile outbreak was purposely planted here, don't rule it out. "As a scientist, you can't—there's just no data," says microbiologist George Ludwig. "This was a natural outbreak," adds Mike Turell. "The question is, did it arrive naturally or did someone assist it? This could be a very subtle, very carefully arranged attack meant to mimic a natural infection."

Regardless of how West Nile virus got here, the big question is: What course will it blaze next? To answer that, investigators must figure out every step of its natural history—just as Ross and Reed did in their signal discoveries about malaria and yellow fever at the turn of the last century. What mosquitoes are carrying it and spreading it? Are they biting people or birds? If birds, which birds? Where's the virus hanging out during the winter? How many dead birds must turn up before human cases emerge? Which birds are sustaining the virus in nature? (House sparrows currently look like a good bet.) And what makes this virus deadlier in birds than other West Nile strains?

"If you want to be a hard-hearted scientist," says UC–Berkeley's Bill Reeves, "this is a grand experiment. The fact that these birds are dying is a beautiful example of what happens when any agent gets into a virgin population." Adds CDC epidemic intelligence service officer Dan Singer, "We're not jumping on this because it's going to wipe out the population of North America. We're keeping a close eye on it because it's the first time we've had a level of technology to see how a disease spreads through a community and the environment. A lot of the public health interest is not West Nile as West Nile, but West Nile as a novel pathogen." Knowing how West Nile spreads may help scientists forecast how Japanese encephalitis viruses or others will behave in a new land.

In 2000, West Nile virus defied all the scientific oddsmakers' predictions. Rather than heading south with the previous year's migration and fizzling out in the Northeast, it wintered comfortably up north and proceeded to carve out a northern and western route, perhaps riding with the spring bird migration, extending as far north as Boston and as far west as the New York–Ohio border, as well as south to North Carolina. Now that it seems here to stay, scientists expect it to fan out into all regions that SLE populates—including South America—on the wings of migratory birds. With one audacious leap, an Old World pathogen became an American virus-in-residence. And it won't be the last.

A Second Look

And that's just what unnerves public health officials. West Nile established itself here, leaving clues literally at New Yorkers' feet, before anybody put the pieces together. When things go wrong like this, officials ruefully haul out the "retrospectoscope": a crystal-clear lens that focuses the mistakes of a recent investigation. Since 1999, it's gotten a lot of use.

Critics have described the West Nile investigation as a "failed dress rehearsal" for a still-evolving national plan to respond to bioterrorism. Getting the virus wrong was just part of the problem. Tunnel vision blinded CDC investigators to evidence that contradicted their original judgment. Scientists neglected to remember that every other SLE outbreak in this country had begun in the south and crept up along the Ohio or Mississippi river valleys—in other words, that there had always been inklings in the south before anything happened up north. They also ignored the fact that muscle weakness—the most striking symptom in the early cases—had never been a hallmark of SLE. They didn't quickly probe the reasons for the ambiguous test results on patients' blood, and dismissed New York City officials' repeated questions about those results. Perhaps most important, they slighted the concerns of Tracey McNamara, the scientist who early on uncovered the central clues to the mystery.

"Our public health system and the response to this crisis came down to a few personalities. And it's frightening that some quirks can obstruct an investigation," McNamara said. And perhaps, she added, a not-so-subtle misogyny also got in the way. The fact that she and the other women who raised early questions do not "speak with a deep baritone was a very important factor. I hate to say it, but that's still the way of the world."

Among the onlookers who was not surprised by this series of mistakes was a former director of the CDC's National Center for Infectious Diseases. Virologist Frederick Murphy, one of the co-discoverers of the Ebola virus who is now at the University of California–Davis, had just a year before prophetically published an article pleading that government officials listen to those outside the "citadel," the traditional federal community of investigators and officials—people such as local clinicians, pathologists, veterinarians, and animal scientists, whose observations can be crucial early gleanings of an emerging epidemic. At the time, Murphy's former federal colleagues castigated him for publishing such a solecism. But today, CDC–Ft. Collins convenes weekly conference calls with all of the 48 mainland states as part of its West Nile follow-up—calls that include public health veterinarians, wildlife experts, and representatives from the United States Department of Agriculture—during which speakers can brainstorm about any suspicious animal-to-human infections.

Another deep-seated problem also hobbled the investigation: the decrepit state of arbovirology. Though the arrival of West Nile virus in the United States would open a rich vein of research, for some researchers it also triggered a kind of professional anguish. "My first reaction was: 'We're in a hell of a lot of trouble,'" said Yale University entomologist Durland Fish. "Because the field of vectorborne disease ecology is almost extinct."

From the 1940s through the 1960s, arbovirology was hot science. After a search for the yellow fever virus uncovered a teeming universe of viruses in the tropics, the Rockefeller Foundation set up a worldwide network of laboratories, from Brazil, Colombia, and Trinidad to India, South Africa, and Uganda. Partly, the goal was to find and catalog hitherto unknown viruses—which the foundation did, by the hundredfold. Partly, it also aimed to nurture new technologies and donate them to the host nations.

"I didn't know it at the time, but it was as good as it gets," says arbovirologist Robert Shope, a granddaddy in the field. "We had a lot of freedom. We were not saddled with writing grant proposals." Back then, arbovirologists were generalists *par excellence*—in addition to virology, they mastered entomology, vector biology, ecology, mammology, ornithology, and epidemiology. CDC–Ft. Collins amassed a huge library of reagents and built a state-of-the-art lab to identify new mosquito species. According to Tom Monath, "People were in tune with global virology. You had a cadre of people who had tremendous international experience." Here at home, CDC scientists explored the natural history of St. Louis encephalitis, Eastern equine encephalitis, and Western equine encephalitis.

By the mid-'60s, however, success began to breed indifference. As vectorborne infectious diseases such as malaria, dengue fever, yellow fever, and plague retreated in most parts of the world except Africa, their breeding sites eliminated with chemical pesticides and other means, interest in the field languished. "When I arrived at Johns Hopkins in 1965," says the CDC's Duane Gubler, "I was told that I was in a dying field."

The 1975 outbreak of SLE in the United States temporarily fired up the field again. State and local health departments launched programs to catch mosquitoes, capture sparrows and test their blood, and perfect the use of so-called sentinel chickens, which were placed in mosquito-ridden areas and tested regularly to see if they had been bitten and produced antibodies to the SLE virus. But when there were no subsequent outbreaks for five years, legislatures shifted the money to other programs. Because vectorborne epidemics come and go without warning, health officials lament, it's always been hard to persuade legislators to keep up the flow of research funds. Along the way, arbovirology became scientifically fragmented. The federal program focused on molecular biology rather than fieldwork. Most state programs, which do the yeoman's work in picking up and containing epidemics, deteriorated. "We let it go too far," admitted Gubler. "Many of the states simply don't have the people or the facilities to deal with these diseases." According to the CDC's Steve Ostroff, who since 1999 has been West Nile coordinator for the U.S. Department of Health and Human Services, West Nile in New York City represented the "chickens coming home to roost."

Today, few training centers or universities are planting the next crop of arbovirologists and medical entomologists. "We need new ideas," says Yale's Durland Fish. "We need new methods for surveillance. We need new methods for control. The light traps they're using is a fifty-year-old technology. Spraying is a one-hundred-year-old technology. That doesn't speak very well for the field. We're almost beyond the point of no return. West Nile came in the nick of time."

***

On June 27, 2000, Flushing Hospital's Deborah Asnis—the woman who arguably tipped off the world to West Nile virus in New York City—phoned Annie Fine at the department of health. Asnis had a patient who couldn't move his arms or legs. For a single suspended moment, it looked like lightning was about to strike twice. But it didn't. This unfortunate victim had dined on rat poison, and not voluntarily. He died the next day.

Just as it takes a series of coincidences for a traveler like West Nile virus to land here and thrive, so noticing it depended on an astute clinician and vigorous surveillance. Public health officials say they can't keep pathogens out—they can only hope to notice them early and act quickly. What if Deborah Asnis had never called the New York City health department? Would the outbreak have been noticed only after it had already subsided, forcing public health officials to go back and pick up the pieces? Would it have been noticed at all? The first weekend Layton and Annie Fine looked for cases, 19 West Nile encephalitis patients already lay in the hospital—but only four had been reported.

What if the outbreak had started somewhere else? "There's 10 million people in metropolitan New York. There were seven deaths," said the CDC's Nick Komar. "Say there's two million people in metropolitan Boston: there would have been one or two deaths—let's say one. If there was one encephalitis death in a hospital in Boston, would they have ever tracked it down to an arbovirus? Maybe not. . . . And if we didn't know about West Nile virus in 1999, what would we be seeing in 2000? I'm guessing that we wouldn't have noticed anything."

"To expect the unexpected" has for years been the mantra of public health scouts on patrol for new and emerging infections—so much so, the phrase runs the risk of fading into background din. The 1998 CDC report *Preventing Emerging Infectious Diseases: A Strategy for the*

21st Century, warns that "because we do not know what diseases will arise, we must always be prepared for the unexpected." In his autobiography, *Virus Hunter,* C. J. Peters offered a similar and much-quoted piece of wisdom from the obverse side: "Common things occur commonly. Uncommon things don't. Therefore, when you hear hoofbeats, think horses, not zebras." For a few weeks late in the summer of 1999, the zebra's stripes were camouflaged by the shadows of dead crows, and all investigators could see was a horse.

The Zoonotic Threat

Though West Nile virus caught the attention of citizens and the press, what wasn't played up was the larger threat that the virus exemplifies. West Nile virus is a zoonotic infection—it jumps to us from animals. It is not the most fearsome zoonosis; but zoonoses have shadowed us throughout human history and remain an ever-present menace, a limitless reservoir of pestilential surprise. To understand what's on the minds of public health officials, it helps to take a quick detour into this territory.

Of the more than 1,700 known viruses and bacteria and other pathogens that infect humans, about half come from animals. Of the more than 150 pathogens linked with emerging infections, about three-quarters come from animals. In the past quarter century, scientists have cornered more than 30 new infectious diseases—and all but a few have come from animals. Whatever deadly pandemic next sweeps the world, whatever newly christened scourge dominates headlines, it will almost surely have jumped species in this way, as measles and influenza once did. There is no telling what it will be, where it will come from, whether it will be an isolated curiosity or a ravaging epidemic—or even whether it will come courtesy of a bioterrorist, since many of the viruses and bacteria most likely to be used in such an attack are zoonotic, such as those causing anthrax, plague, brucellosis, and tularemia, as well as equine viruses.

A zoonosis emerges when an agent gains access to a new place or new conditions that promote its spread. The Ebola virus, a filovirus whose natural host in the wild is unknown, was discovered in 1976 and originally transmitted in jungle and savannah areas where humans had newly encroached; a Zaire outbreak killed 90 percent of its victims. Marburg virus, another filovirus that causes the massive internal bleeding of hemorrhagic fever, was discovered in 1967 in German and Yugoslavian lab workers and vets who handled tissues from African green monkeys. The Nipah virus emerged in 1998, a virus harbored in fruit bats that killed 40 percent of slaughterhouse workers who caught it from pigs and developed encephalitis. Zoonotic infections reach us either directly from animals, or indirectly from animals via insect vector, such as West Nile virus.

For early humans, progress invited zoonotic infections. When they descended from the trees roughly five million years ago, our ancestors picked up parasitic worms from the flesh and feces of animals on the savannah and sleeping sickness from the tsetse flies that preyed on those herds. Meat-eating, tool-wielding *Homo erectus* set out from Africa to the warmer parts of Europe and Asia one million years ago and traded the chronic illnesses fostered by tropical parasites for a new crop of afflictions. In *Man and Microbes*, Arno Karlen surveys these new torments: sporadic cases of bubonic plague (transmitted by rats); botulism and staphylococcus infections (from scavenged meat); gangrene and tetanus (from butchering kills); relapsing fever, hemorrhagic fevers, brucellosis, leptospirosis, toxoplasmosis, and salmonellosis (transmitted by wild game); scrub typhus and encephalitis (from insects and ticks).

Not until around 3000 BC, when agriculture had given rise to larger settlements, did what historian William H. McNeill calls the "diseases of civilization par excellence" begin to take hold. These diseases are acute, travel swiftly and efficiently, and tend to be limited to humans. Biologists believe that the measles virus stemmed from canine distemper and rinderpest, an affliction of cattle; that rhinoviruses—agents of the common cold—came to us from horses; that the human form of tuberculosis derived from a mutation in the bovine strain; and that smallpox is a close cousin of cowpox. As Jared Diamond points out in his book *Guns, Germs, and Steel*, "when we domesticated social animals, such as cows and pigs, they were already afflicted by epidemic diseases just waiting to be transferred to us."

Zoonotic agents spread rapidly when they become biologically comfortable enough in *Homo sapiens* to make us their new home. The most frightening launchpads for emerging zoonoses are nonhuman primates, since viruses that afflict monkeys and apes are most likely to adapt to us. The most shocking example, of course, is the AIDS virus, first discovered in 1981, which continues to evolve and expand. In 2000, it newly infected more than three million people. The AIDS global epicenter is sub-Saharan Africa, where two versions of the human immunodeficiency virus leapt the species barrier: HIV-1, which originated in chimpanzees, and HIV-2, which came from sooty mangabeys. The dominant virus in the global epidemic, HIV-1 seeded itself in humans in the early part of the twentieth century, conceivably when hunters killed and dressed wild chimps and ate their raw meat, a common practice in west central Africa, where the animals' infected blood infected people through cuts or through mucous membranes. Sooty mangabeys, which carry HIV-2, were not only hunted for food but also kept as pets. According to a genetic family tree drawn by the world's fastest computer—the Nirvana at Los Alamos National Laboratory—HIV first jumped into humans around 1930. In 1985, scientists from Emory and Harvard Universities found evidence of the AIDS virus in a preserved specimen from an unidentified African man who died in 1959 near present-day Kinshasa, capital of the Democratic Republic of Congo. Beatrice Hahn, a physician/researcher at the University of Alabama at Birmingham, believes at least 24 other African primate species are infected with similar retroviruses. Hahn worries that new commercial logging operations in the tropical forest will foster a "bushmeat" trade—the large-scale hunting and consumption of wild animals, traditionally a subsistence activity. With newly cleared paths of transmission, new human-adapted primate infections could come our way.

Certain conditions prime the pump for animal-to-human infections. Untouched animal and arthropod habitats—from tropical rainforests to American woodlands—are disease minefields for humans. Though travel and transport thrive more in the developed world, ecological change is most rampant in the developing world, and biodiversity richest in the tropics—making the latter the best hunting ground for emerging infectious agents. Megacities of the tropics, with their crowding and poor sanitation, act as incubators for emerging zoonoses, and porous public health surveillance gives them a free pass. Nearly all of the world's population growth in the next 25 years will occur in urban areas of developing countries. But disease-incubating conditions aren't confined to what we think of as the third world. Bucharest in 1996 saw an epidemic of encephalitis caused by West Nile virus in an area where apartment building basements were flooded, creating such massive mosquito populations that the insects blackened the walls and ceilings of apartment buildings. Global transportation further stirs the pot, delivering disease agents to new ecological milieus where they can prove their adaptive mettle. "If you look twenty or thirty years down the road," says Duane Gubler, "all of the factors that have been responsible for the resurgence of these diseases are going to get worse before they get better."

In the coming century, climate changes may fuel more vectorborne outbreaks. According to the 2000 United States National Assessment Report, an analysis from experts in academia, government, and private industry, the U.S. will see higher temperatures, more extreme weather shifts, and a different precipitation cycle over the next hundred years. Because vectorborne disease cycles are exquisitely tuned to the environment, the effect of such shifts, whether we will see more or fewer flareups of West Nile encephalitis or other vectorborne diseases, isn't clear.

Some scientists are skeptical that global warming will unleash a torrent of vectorborne infections. For one thing, notes CDC entomologist Paul Reiter, during the coldest period of the Little Ice Age—from about 1564 to the 1730s—malaria was rife in England. Indeed, during the eighteenth century, the disease reached as high as the Arctic circle.

All this points up the importance of nonclimate factors, such as drainage and land reclamation, the decline of rural populations, mosquito-proof houses, and the capacity to raise more livestock, which in the case of malaria may have diverted biting mosquitoes from human flesh. Nonclimate factors may also protect us today. While dengue has flared in Latin America, for instance, the infection screeches to a halt at the U.S. border. From 1980 to 1996, there were more than 50,000 cases of dengue in the three Mexican states contiguous to Texas—and only 43 in Texas itself. That sudden shift in disease at the border may be partly attributed to the displacement of *Aedes aegypti* by *Aedes albopictus*, a poor vector for arboviruses to humans because it is as apt to bite rats, cats, and dogs as people. But more critical has been the U.S. standard of living and our social (or antisocial) habits. "We don't go out in the evening or the afternoon and talk to our neighbors," says Gubler. "We're in, watching television in air-conditioned houses that are hermetically sealed."

But other scientists contend that as the atmosphere warms, glaciers will melt, sea levels will rise, and storms will become more uninhibited. If temperatures at night and in winter rise, it will speed the reproduction of viruses and parasites. Rising temperatures will also boost the metabolism of insects and other arthropods, forcing them to feed more and thus putting more people at risk for West Nile encephalitis, malaria, and other diseases.

The 1993 hantavirus outbreak in the American Southwest may have had roots in an unusual weather pattern, similar to that prompted by global warming, known as the El Nino/Southern Oscillation. An ancient strain that first caught health officials' attention in 1993, when it killed more than two dozen people in the Four Corners region of the rural Southwest, it is inhaled in the aerosolized, dustborne wastes of the deer mouse. More than 40 percent of victims die of an infection that first feels like the flu but ultimately turns the immune system into a self-annihilating weapon, destroying capillary walls in the lungs and filling the lungs with fluid. In 1993, after six years of drought, heavy spring rains pelted the region, leading to a tenfold increase in the population of deer mice, the animal reservoir for the Sin Nombre virus.

Writing in *Science*, Harvard University physician and climate change researcher Paul Epstein invokes global warming as the stage for New York City's West Nile outbreak:

> The mild winter of 1998–99 enabled many of the mosquitoes to survive into the spring, which arrived early. Drought in spring and summer concentrated nourishing organic matter in their breeding areas and simultaneously killed off mosquito predators, such as lacewings and ladybugs, that would otherwise have helped limit mosquito populations. Drought would also have led birds to congregate more, as they shared fewer and smaller watering holes, many of which were frequented, naturally, by mosquitoes. Once mosquitoes acquire the virus, the heat wave that accompanied the drought would speed up viral maturation inside the insects. Consequently, as infected mosquitoes sought blood meals, they could spread the virus to birds at a rapid clip. As bird after bird became infected, so did more mosquitoes, which ultimately fanned out to infect human beings. Torrential rains toward the end of August provided new puddles for the breeding of *C. pipiens* and other mosquitoes, unleashing an added crop of potential virus carriers.

***

The same year that West Nile virus hit New York City, dozens of less-publicized pathogens leaped from animals to people every day in every part of the United States. Though not all are important from a public health standpoint, they illustrate a timeless exchange between humans and our planet's cohabitants. Just as West Nile virus took advantage of human activities to annex new territory, so have local pathogens.

To get a sense of how imperceptibly our paths cross those of pathogenic bacteria and viruses, consider an outbreak that struck in the heart of the Midwest a year before West Nile emerged in New York. Just after dawn on Father's Day, 1998, 851 men and women, attired in swimsuits and goggles, gathered at the edges of Lake Springfield for the start of the 16th Annual Iron Horse Triathlon. The heavy rains of the previous day and night had cleared. It was sunny and familiarly

humid for the middle of Illinois in summer. In timed waves, the con-
testants launched an assault across the flat midwestern landscape, first
swimming 1.5 miles, then biking 45 miles on country roads edged by
corn and beans, finally running ten miles on sticky two-lane blacktop.
It was hardly an environment that suggested disease.

Within two weeks of the competition, 97 of these athletes were
sick with wide-ranging but eerily similar symptoms. Abruptly and without warning, they had developed fever, chills, muscle pain, and headache. Some went on to suffer more severe GI and neurological problems. Twenty-three were hospitalized.

Several years earlier, these same symptoms had turned up in a very
different group of patients: residents of inner-city Baltimore. One, a
young woman, had gashed her foot while walking shoeless in an alley
where she had seen rats. A man recalled accidentally piercing his hand
on glass in a rat-infested passageway between row houses. Another
man wandered barefoot near Baltimore's Inner Harbor.

What connected these mysterious outbreaks in such disparate locales and groups of patients? A tiny, slender, helically shaped bacterium of the species *Leptospira interrogans.* The signature disease caused
by this spirochete is leptospirosis, thought to be the most widespread
zoonotic disease on the globe. Workers in rice fields, sugar cane plan-
tations, and mines are highly susceptible. Each year, an estimated 100
to 200 Americans contract leptospirosis, though the number may in
fact be far higher. People get leptospirosis from fresh water, wet soil,
or vegetation contaminated by the urine of infected animals. The bacterium enters the body through broken skin, mucous membranes in
the eyes and nose, or contaminated drinking water. Lake Springfield
had likely become contaminated when heavy rains washed animal
wastes into the lake, which was then churned by the unwitting
triathletes. The Baltimore patients caught the disease by inadvertently
touching the urine of city rats, which are known to harbor the bacterium. In fact, some scientists assert that leptospirosis, a disease that
contributed to Napoleon's defeat in Poland and Russia during the winter of 1812, goes vastly underdiag-nosed in this country's urban patients. One recent study found that of patients whose blood was drawn

at a Baltimore clinic for sexually transmitted diseases, 16 percent had been exposed to *Leptospira.* A study from Detroit found that inner-city children were far more likely to harbor antibodies to the bacterium than suburban kids—and no one knows the consequences of these untreated infections.

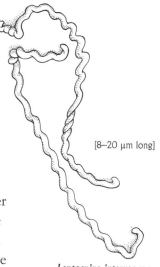

[8–20 μm long]

Leptospira interrogans
Agent of leptospirosis

But leptospirosis pales beside other home-grown zoonoses. Each year, more than 16,000 Americans come down with the fever, fatigue, and rash of Lyme disease, a bacterial infection conveyed by the bite of the deer tick. Untreated, the infection can progress to arthritis, neurological problems, and heart disease. *Borrelia burgdorferi* spends its life on the white-footed mouse and drinks blood from the white-tailed deer, denizens of the modern suburban and exurban checkerboard of vegetation. Here, small, fragmented woodlots and tall grasses have blossomed on forest land originally cleared in the nineteenth century for agriculture, and these species have exploded in number. Lyme disease didn't come to us; we came to Lyme disease, by intruding on a tilted ecosystem of our own making. Since 1982, when surveillance began, the number of annual cases has increased 25-fold. According to recent studies, forest tracts unbroken by asphalt and human abodes sustain a wider array of small mammals and birds than do today's carved-up sylvan fragments. In these larger forest expanses, where white-footed mice face more competitors and more predators, fewer ticks carry the Lyme bacterium.

In 1999, Americans were sickened with Sin Nombre virus; bubonic plague, carried by the fleas of prairie dogs, squirrels, and rodents; ehrlichiosis, a burgeoning zoonotic disease caused by a rickettsial organism carried by the same tick that transmits Lyme disease;

tickborne Rocky Mountain spotted fever and babesiosis; mosquito-borne LaCrosse encephalitis and Eastern equine encephalitis; bacterial psittacosis from parrots and *Salmonella* infections from iguanas, and many other animalborne diseases—including the millions of foodborne infections from livestock that top the list of zoonoses. New York City, the world's preeminent urban milieu, is not immune to seemingly rusticated ambushes. It has seen cases of tularemia, a rare human infection often found in rabbits, cats, and voles, and Rocky Mountain spotted fever, a tickborne infection that despite its name has been found in almost every state in the country.

If disease cops had a handle on every zoonotic assault that takes place in the United States, the daily police blotter would be as long as a big-city phone book.

Race to the Finish

Though readers got blow-by-blow reports of the public health establishment's duel with West Nile virus, what they didn't read about was the behind-the-scenes race to identify the virus. This story is like that of many highly publicized investigations. On Saturday morning, the day after the CDC's bombshell announcement, the agency's Rob Lanciotti, John Roehrig, and Duane Gubler were talking in Gubler's office when the phone rang. It was Ian Lipkin, a scientist at the University of California–Irvine. Unbeknownst to Lanciotti (and to the New York City health department), Lipkin had also been pulling all-nighters to pinpoint the pathogen killing New Yorkers. A neurologist by training, he directed the university's Emerging Diseases Laboratory. Beginning in the late 1980s, Lipkin had made a name for himself by exploring the hidden infectious causes of neurological diseases, along the way developing innovative molecular biology tools for ferreting out pathogens. In Lipkin's lab, DNA searches that used to take days now took only hours.

At a scientific conference about encephalitis testing in September, officials at the New York State Department of Health laboratory told

Lipkin that they were having trouble amplifying genetic sequences of the SLE virus from the tissues of New York City's encephalitis victims. Lipkin, a trim, precise, almost elegant man with wire-rimmed glasses and an aquiline nose, leaped at the chance to put his innovative research methods to the test. By Wednesday evening, September 22, he had found the genetic footprints of a flavivirus in the encephalitis victims. By Thursday, he had unspooled pieces of the virus's genetic sequence. In doing so, Lipkin was the first to close the loop in an epidemic involving humans, birds, and mosquitoes. The CDC called the virus "West Nile–like"; Lipkin called it "Kunjin/West Nile–like," after a strain of West Nile virus found in Australia. Because the taxonomy of Kunjin is somewhat ambiguous, both teams were right. The discrepancy arose because Lipkin and the CDC's Lanciotti were comparing different sections of the genome to different reference strains in GenBank.

When Lipkin called Gubler's office that Saturday morning, Gubler and his colleages were expecting a technical ironing-out of their divergent results in this high-profile outbreak. Lipkin had something else in mind: a quick joint publication in the *New England Journal of Medicine*. "I said, 'Look, you've got the data in the animals. We've got the data in the humans. In concert, we've established that this is an important new zoonosis. We should write this thing up together.'" The CDC scientists were silent. Finally, Gubler—a big man with a thick gray moustache, cat eyes, and the appearance of a TV Western frontiersman—spoke up. Gubler was himself an aggressive scientist. But the CDC, he said, was not interested in publishing until the agency had rock-solid data about the virus. For good or ill, any prospective public–private collaboration ended there.

Suddenly, the endeavor to fully sequence the virus was a competition. "It became absolutely clear to me that these guys were going to steal credit for everything that had been done," Lipkin said. "Because basically, without the human work, they wouldn't know what they had." On October 9, Lipkin published his Kunjin/West Nile–like findings in the British medical journal *Lancet*, seemingly beating the CDC to the finish line. But in the end, it was the CDC that hit the bull's eye in an announcement two weeks later: New York's virus was most closely related to an unpublished 1998 West Nile sequence from Israel, which the CDC had acquired from a French scientist—a sequence to which only the CDC had full access.

Many onlookers praise Lipkin for first discrediting the St. Louis encephalitis diagnosis. "You've got to give him credit," says Tom Monath. "He looked at the sequence and recognized this was West Nile–like virus and snapped off a paper. Well, that's the world we live in—especially the academic world." Others were angry with Lipkin for rushing to publication with what the CDC still claims are misleading results.

When egos are roused, it's hard to get at the truth. Lipkin says that in spurning cooperation, the CDC not only set a bad precedent but also set back science. "They're bulls in china shops. And I guarantee you that the next time something comes around that's a little bit squirrely, they're going to miss it again. . . . My view is that you can get a lot more done working with people than fighting. And life is short." CDC officials publicly affect a cool respect toward the academician, while suggesting between the lines that he was something of an irritant. The competition left a bad taste for many onlookers and brought up a larger issue: In a public health emergency, can public health and academic researchers collaborate? Clearly, the stakes for each are different. Public health investigators need to solve a crisis, academics to garner recognition and funding. Within the CDC, many branch directors believe they wouldn't be able to keep their best bench scientists if they didn't protect those researchers by making sure they were the first to publish new findings. But others believe such protection is unnecessary, and that in any case CDC should rise above proprietary concerns. "In the good old days, CDC didn't compete," says Frederick Murphy. "The motivation was driven by the sense that CDC's unique, that it's a public service organization—and that the reputation of the scientists will prosper anyway, because of the incredible opportunity that the CDC has to get all this good stuff."

When a historic outbreak strikes, "everyone wants a piece of it," Duane Gubler said. "You know, if you're in science, you don't get paid a lot. Your rewards are basically building up your ego, right? And the way you build up your ego is to do good science and write publications and be the first to discover things. That's what drives it. And people's memories get clouded because of that."

So were memories clouded? Or is there something about fast-breaking, high-stakes science that makes the protagonists the heroes of their own dramas? Looking back, New York City's Marci Layton was intrigued by what she calls the "Rashomon quality" of the investigation. Deborah Asnis, Tracey McNamara, Ian Lipkin, Rob Lanciotti—each could plausibly claim to have "discovered" West Nile virus in the Western Hemisphere. The question is: Who will get to write the history books?

Epilogue

In 1999, 62 people were hospitalized with West Nile virus, seven of whom died, including a businessman who had visited New York City from Toronto and had become ill after he returned home. Most patients had become ill during the third and fourth weeks of August. While the greatest concentration of cases was in northern Queens and the south Bronx, investigators also found the virus in birds and mosquitoes in eastern Long Island, the lower Hudson Valley, New Jersey, and Connecticut, as well as in 25 horses (of which nine died) in Suffolk County, New York, and in a cat in New Jersey. New York City's West Nile hotline logged more than 150,000 calls. In late September, Greenwich, Connecticut, officials temporarily banned "absolutely all outdoor activities" after 5 p.m. In New York state alone, residents reported 17,000 dead birds. Approximately 10,000 crows died in the outbreak. Researchers collected 25,000 mosquitoes for testing. As part of its effort to wipe out mosquito breeding sites, New York City erased one of its aesthetic signatures: abandoned tires.

In October 1999, Annie Fine got married in Prospect Park, at a ceremony that took place at dusk, when *Culex* mosquitoes like to bite. She worried about her 96-year-old aunt. The toasts were replete with West Nile jokes. In a footnote worthy of Edgar Allan Poe, a solitary dead crow infected with the virus turned up in faraway Baltimore. The European Union banned horse and poultry imports from affected areas in the United States. That same month, investigators drew blood from Queens residents to find out how many people had been silently infected during the epidemic. They estimated that 2.6 percent of residents, or about 8,200 individuals, had West Nile virus antibodies in their blood; most had experienced no symptoms at all.

In January and February 2000, investigators gathered mosquitoes from sewers near the Bronx Zoo, from under the Whitestone Bridge, and from underground rooms at Ft. Totten. Three Ft. Totten samples yielded virus-positive mosquitoes, meaning the virus would return in the spring. And it did. Federal, state, and city agencies resumed their surveillance for infected birds and mosquitoes, spraying pesticides wherever the virus turned up. That year, for the first time in memory, there was a sudden run on arbovirologists, medical entomologists, and mosquito control experts. In April, New York City began its campaign to kill mosquito larvae, dropping poisonous pellets in the city's 150,000 storm sewers. In every borough, white leghorn chickens deputized by the health department sat in secretly sited cages exposed to biting mosquitoes all spring and summer, bled each week by a veterinary technician to see if they had developed antibodies. All summer long, cities up and down the eastern seaboard, along the Gulf Coast, and inland as well searched for signs of West Nile virus. Eventually, the pathogen had spread to 12 eastern states and Washington, D.C., sickening 21 and killing two.

In 2001, the virus struck humans and animals in the Deep South and the Midwest. Experts predicted that by 2006 the entire nation would play host to West Nile virus, and that additional species of daytime biting mosquitoes would transmit infection. Dead birds remained the most sensitive way to track the virus in real time over a widening geographic area. Daily newspapers ran front-page bird obituaries. Inevitably, West Nile shock turned into West Nile shtick. A Westchester

County columnist surmised that it was not the virus, but the name of the virus that stirred fear. Had it been called "West Nyack virus," he wrote, nobody but real estate agents would care. Two years after his scientific coup, Ian Lipkin decamped from UC–Irvine to Columbia University. And today, when she talks to scientific groups about her trials and travails in the West Nile virus investigation, Tracey McNamara never fails to flash on the screen a slide of two bucks locking antlers. "I work in a zoo," she quips. "I recognize territorial display behavior."

Duane Gubler, the Utah rancher's son who spent 21 years outside the United States in tropical countries, felt a bit rueful. He wondered why Americans failed to grasp West Nile's transcendent message. Why didn't they see the connection between a stirring little domestic drama and what happens every day far beyond their borders? "Most people who suffer from these diseases are in tropical countries," he said. "What makes me a little bit sad—not a little bit, but just *plain* sad—is that you've got diseases that kill tens of thousands of children in developing countries, and nobody cares about that until it gets to home."

Chapter 3

Food Fright

In the autumn of 1996, Laurie Girand, a marketing consultant in Silicon Valley, came home from a Fiji vacation to find her three-year-old daughter Anna in constant stomach pain. "My tummy hurts. My tummy hurts," the usually resilient child kept saying. Within days, Anna was in the hospital, severely anemic, her red blood cells looking like shredded circles under the microscope. Doctors diagnosed her with hemolytic uremic syndrome, or HUS, a sometimes fatal complication of foodborne infection. The bacterium behind the disease, *E. coli* O157:H7, is usually linked to undercooked hamburgers—Girand vaguely remembered a huge Jack in the Box outbreak of a few years earlier. But her daughter hadn't gotten sick from eating hamburgers. Only as Anna lay in a hospital bed, ashen-faced, glassy-eyed, waiting for a blood transfusion, did Girand and her husband hear a news account that apple juice had been linked to an *E. coli* epidemic. Manufactured by Odwalla, Inc., its selling point was that it was unpasteurized, and presumably more wholesome. The apple juice had been a treat from Anna's grandmother while Girand and her husband were away. And though Girand didn't approve of apple juice—too many empty calories—for a long time she had been feeding her daughter Odwalla unpasteurized carrot juice. "I was under the seriously mistaken impression that feeding our daughter unpasteurized juice would be healthier for her," Girand says. "Odwalla's slogan at the time was: 'Drink it and thrive.'"

Though anemic for months, Anna recovered. But in Colorado, a child had died, while nearly 70 others, mostly children six and younger, had become severely ill in what was, until then, the country's biggest juice-associated outbreak. Catalyzed by the near-tragedy, Laurie Girand started giving speeches to parents and writing letters to government officials. But shocking as the Odwalla outbreak was, it was not sufficiently instructive. Three years later, in the nation's biggest juice outbreak, *Salmonella* in unpasteurized Sun Orchard orange juice struck nearly 500 victims and killed one. A few months later, the company had to recall another tainted lot. "I can't believe," Girand says, "since I am a marketing person, how badly I was fooled by industry marketing."

One of the most insistent marketing messages we hear, trumpeted by both industry and regulators, is that the United States has the safest food supply in the world. Yet according to the CDC's best calculations, each year 76 million Americans—nearly one in four, and that's a lowball estimate—become infected by what they eat. Most find themselves for a few days dolefully memorizing a pattern of bathroom floor tiles. About 325,000 land in the hospital. Two million suffer drawn-out, sometimes lifelong medical complications from unwittingly eating a contaminated morsel. More than 5,000—about 14 a day—die from indulging in what should be one of life's great pleasures. The "world's safest food supply" regularly doles out *E. coli* O157:H7 in hamburgers, *Salmonella* in alfalfa sprouts, *Listeria* in hot dogs, *Campylobacter* in Thanksgiving turkeys.

Change is what ushers new disease-causing organisms into our lives. And in the past few decades, there have been profound shifts in what we eat, where our food comes from, how it's made, and who makes it. Fifty years ago, grocery stores stocked about 200 items, 70 percent of which were grown, produced, or processed within a 100-mile radius of the store. Today, the average supermarket carries nearly 50,000 food items, some stores as many as 70,000. Agriculture and food manufacture have grown into global economies of scale, producing megaton quantities that, if contaminated, increase the potential for widespread epidemics. More fresh fruits and vegetables come from abroad, where sanitary standards may not be as high as in the United States. And our meals are increasingly cooked by people untrained in the techniques of safe food preparation.

This is not your grandparents' "food poisoning"—a now-quaint term that originated early in the twentieth century, when dramatic gastrointestinal distress was usually traced to toxins, especially staph toxins, that had grown on spoiled foods such as cream-filled pastries or chicken salads left out too long in summer heat. Literally cases of food "intoxication," these infections struck suddenly and fiercely, usually within two to six hours after the meal. When local health officials worked up these classic "point source outbreaks," they would inevitably find that a knot of victims had all eaten a single dish, and that cases sharply climbed and then plummeted as the well of exposed individuals dried up. Point source outbreaks haven't faded away; big-city health departments face dozens every year. In 1997, for instance, *Salmonella*-tainted hams from a church fundraising dinner in St. Mary's County, Maryland, sickened 700 people and killed an elderly woman. Today, however, the modest church picnic has given way to a giant food bazaar created by massive consolidation and global distribution. One contaminated tidbit—a shred of meat from an infected steer mixed with hundreds of other carcasses for hamburger, an iced box of tainted lettuce dripping down on the rest of an outbound lot, a soiled production line of cereal shipped coast-to-coast under 30 different brand names—spreads disease far and wide.

The pathogens in science's crosshairs have also changed—in part because improved technology permits scientists to see some for the first time, and in part because evolution has selected for more noxious creatures. Twenty years ago, today's most fearsome threats were overlooked or yet-to-be-discovered. *Campylobacter jejuni*, now known to be the most common bacterial agent in food, was considered a rare, opportunistic organism because lab workers didn't see it hiding among less fastidious bacteria growing in culture. A small, delicate, spiral-shaped microbe, it corkscrews its way into mucous membranes of the intestinal tract "with a speed that cannot be matched by other bacteria," according to one scientist's report. *Listeria monocytogenes*, the most deadly agent in our food supply, killing one in five victims it infects, wasn't even suspected of spreading through food. *E. coli* O157:H7, a potent threat to children and the aged, was identified only in 1982—and even then remained a medical curiosity until the infamous 1993 Jack in the Box hamburger outbreak. Norwalk virus, the top cause of foodborne illness in this country at 23 million cases a year, remained largely elusive until molecular tests revealed it in the 1990s. All of which suggests there are novel disease-causing agents still hiding incognito in our food. Even with modern diagnostic tools, in 81 percent of foodborne illnesses and 64 percent of deaths, doctors don't know what organisms to blame—in part because they don't know what organisms to look for.

To doctors and scientists, some of these bugs—particularly *E. coli* O157:H7—are scarier than anything seen before. "Foodborne pathogens are not purely a bit of nausea and vomiting and diarrhea," says David Acheson, an *E. coli* researcher at Tufts University School of Medicine. "They can kill a previously healthy person in the space of a week." Evolutionary biologists fear that our efforts to eliminate pathogens on the farm and in processing—by, for example, using disinfectant rinses—may paradoxically help select for more durable and virulent strains.

Meanwhile, more of us are more vulnerable to foodborne microbes. Individuals with impaired immunity—the very young, the very old, and people with cancer, organ transplants, diabetes, AIDS, and other conditions that weaken the body's defenses; all told, about a quarter of the population—are more apt to succumb to these infections. Men and women over 65, who in the next three decades will make up one-fifth of the population, produce less acid in their stomachs, eliminating the first line of defense against enteric pathogens; federal officials predict that the aging population could increase foodborne illness by 10 percent in the next decade. Americans are popping more prescription and over-the-counter antacids than ever, and in so doing, giving pathogens entrée to the nether regions of our digestive system where they do the most damage.

Depending on the organism, the palette of symptoms associated with foodborne disease can include diarrhea, cramps, fever, nausea, and vomiting (the notable exception is *Listeria*, which can cause miscarriage, meningitis, and other nonabdominal problems). But that's just the beginning. In some people, researchers have discovered, the gastrointestinal distress that comes and goes with a foul meal may hang around in another form much longer. *Salmonella* can trigger reactive arthritis, an acute joint inflammation. *Campylobacter jejuni* may cause as many as 40 percent of cases of Guillain-Barré syndrome, a severe neurological disorder that can bring temporary paralysis and long-term nerve damage. Other complications include thyroid disease, inflammatory bowel disease, and, should someone survive the struggle against *E. coli* O157:H7, permanent kidney damage from hemolytic uremic syndrome. In these cases, contaminated food seems to provoke an uncontrolled autoimmune reaction. Up to 3 percent of foodborne disease victims—an enormous number, given the total caseload—may suffer lifelong physical problems.

Any depiction of emerging foodborne infections is necessarily panoramic, complex, and accompanied by more questions than answers. This discussion is no exception. As you will discover, debates about questions of farm management, government regulation, and individual versus institutional responsibility may elicit two—or three or four— diametrically opposed arguments that all seem persuasive. "Foodborne illness is more complex than people understand. The more I learn, the less I realize I ever knew," says Mike Osterholm, a former Minnesota state epidemiologist who has probably launched more successful food outbreak investigations than any other public health official in history. "The very nature of the ever-growing and complex food supply chain, and the desire of consumers to have many different kinds of foods available at a moment's notice, has allowed for a whole new spectrum of pathogens to arrive on the scene." What's more, says Osterholm, DNA fingerprinting has pulled back the covers from foodborne outbreaks, showing that "many of the old conclusions we had drawn about what was happening are not valid."

For those of us who don't think about it for a living, it's easy to underestimate the risk of falling ill from food, since the problem is largely invisible—hidden, one supposes, behind the bathroom door. CDC epidemiologists have factored in this cultural aversion by using numerical multipliers that translate the relatively few cases reported into a far higher and more accurate count of victims who never see a doctor. For instance, for every person known to suffer an infection caused by *Campylobacter* or *Salmonella* or *Cyclospora*, there are 38 who have eluded the net of public health officials; for every confirmed case of *E. coli* O157:H7, there are 13 to 27 doubled-over victims. Keep this in mind when you read news stories about foodborne epidemics. The scores of confirmed cases mentioned in wire service stories may actually represent hundreds or thousands of silent sufferers.

Foodborne infections are ubiquitous, sneaky, and regularly sold short. At the CDC, the foodborne and diarrheal diseases branch investigates more outbreaks than any other group in the agency. According to Paul Mead, a medical epidemiologist, "The paradox of foodborne illness is that, on a per meal basis, it's extremely rare. It's like getting hit by a meteor." But in the very act of eating, says Mead, "You're standing in a meteor shower three times a day from the time you're weaned until you die."

Secret Agent O157: The Evolution of a Killer

Every pathogen has a story, but the biography of *E. coli* O157:H7 is especially instructive because it shows how chance favors the prepared germ—and how we are giving certain disease-causing organisms more chances than a rigged roulette wheel. Though *E. coli* O157:H7 has turned up in unpasteurized apple cider in 1991, 1996, and nearly every year since the Odwalla outbreak, it is best known as the agent behind "hamburger disease." Hamburgers, in fact, are Rolls-Royce conveyances for O157. Think of your next Big Mac as the end product of a vast on-the-hoof assembly line. The story begins on hundreds of feedlots in different states and foreign countries. The animals are shuttled to slaughterhouses, where they become carcasses. The carcasses go to plants that separate meat from bone. The boning plants ship giant bins of meat to hamburger-making plants. The hamburger-making plants combine meat from many different bins to make raw hamburgers. At this point, your burger is more fluid than solid, because ground beef continually mixes and flows as it's made, its original ingredients indistinguishable. Grinding also multiplies surface area, so that the meat becomes a kind of soup or lab medium for bacteria. Finally, from the hamburger-making plants, these mongrel patties are frozen and sent to restaurants. A single patty may mingle the meat of a hundred different animals from four different countries. Or, looked at from another perspective, a single contaminated carcass shredded for hamburger can pollute eight tons of finished ground beef. Finding the source of contamination becomes impossibly daunting. (Making juice is also like making hamburgers: one bad apple can ruin a huge batch.) In the Jack in the Box outbreak, investigators found that the ground beef from the most likely supplier contained meat from 443 different cattle that had come from farms and auction in six states via five slaughterhouses. As the meat industry consolidates and the size of ground beef lots grows, a single carcass may have even more deadly potential. In 1997, Hudson Foods was forced to recall 25 million pounds of ground beef for this very reason: a small part of one day's contaminated beef lot was mistakenly mixed with the next day's, vastly spreading the risk.

E. coli O157:H7, the organism that this endless mixing amplifies, is a quiet tenant in the intestines of the 50 percent or so of feedlot cattle it infects, but a vicious hooligan in the human gut. In the bowel, *Escherichia coli*, rod-shaped bacteria first described by German pediatrician Theodore Escherich in 1885, perform a vital task by keeping disease-causing bacteria from taking over. For many decades, that knowledge obscured the fact that some forms of *E. coli* trigger violent disease. *E. coli* O157:H7 (the letters and numbers refer to immune system–provoking antigens on the body and on the whiplike flagella of the organism) was discovered in 1982, during an epidemic spread by undercooked patties from McDonald's restaurants in Oregon and Michigan. The outbreak wasn't highly publicized; even some scientists perceived O157 as more of an academic curiosity than a harbinger of bad things. Eleven years later, the Jack in the Box hamburger chain promoted its "Monster Burgers" with the tag line: "So good it's scary." These large, too-lightly-grilled patties killed four children and sickened more than 700 people—bringing the exotic-sounding bacterium out of the lab and into public consciousness. In fact, however, by the time of the Jack in the Box tragedy, 22 outbreaks of *E. coli* O157:H7, killing 35 people, had already been documented in the United States. Suddenly, fast food hamburgers—a staple of American culture—were potentially lethal.

What makes *E. coli* O157:H7 so fearsome is the poison it churns out—the third most deadly bacterial toxin, after those causing tetanus and botulism. Known as a Shiga toxin, because it is virtually identical to the toxin produced by *Shigella dysenteriae* type 1, it is a major killer in developing nations. The distinctive symptoms of *E. coli* O157:H7 are bloody diarrhea and fierce abdominal cramps; many victims say it's the worst pain they ever suffered, comparing it to a hot poker searing their insides. Between 2 and 7 percent of patients— mostly young children and the elderly— develop hemolytic uremic syndrome, which can lead to death. HUS sets in

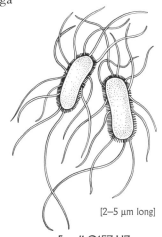

[2–5 μm long]

E. coli O157:H7
Agent of hemolytic
uremic syndrome

when Shiga toxins ravage the cells lining the intestines. The bleeding
that ensues permits the toxins to stream into the circulatory system,
setting up a cascade of damage similar to that of rattlesnake venom.
The toxins tear apart red blood cells and platelets, leaving the victim
vulnerable to brain hemorrhaging and uncontrolled bleeding. Clots
form in the bloodstream, blocking the tiny blood vessels around the
kidneys, the middle layer of the heart, and the brain. As the kidneys
give out, the body swells with excess waste fluids. Complications ripple
through all major organ systems, leading to strokes, blindness, epi-
lepsy, paralysis, and heart failure. Though doctors can manage HUS
symptoms, and are working on new ways to stymie the toxin, they
currently can offer no cure or even effective treatment.

For public health officials, the emergence of *E. coli* O157:H7 is an
object lesson in how a new pathogen can lie low in the environment,
biding its time until humankind changes a certain activity and in so
doing rolls out a red carpet. Like other emerging pathogens, such as
the AIDS virus, O157 had struck long before it caught the attention of
public health officials. In 1955, a Swiss pediatrician in a dairy farm
area first described HUS, which physicians today consider to be a
gauge of *E. coli* O157:H7 infection. Over the ensuing years, the number of cases kept rising, suggesting that O157 was quietly spreading. In
1975, doctors took a stool sample from a middle-aged California
woman with bloody diarrhea, cultured the apparently rare bacterium
and sent it to the CDC, where it sat in storage until the McDonald's
outbreak prompted researchers to scour their records for earlier evidence of the vicious organism. In other words, for nearly 30 years before the first bona fide epidemic, *E. coli* O157:H7 had turned up in
scattered, sporadic cases of bloody diarrhea. It was out in the meat
supply, but not in high enough concentrations to catch health officials'
notice.

Where did *E. coli* O157:H7 come from in the first place? Scientists have pieced together a long, rather provocative history. Genetic
lineages suggest that about 50,000 years ago, O157 and another closely
related serotype—O55:H7, which causes infant diarrhea in developing nations—split off from the same mother cell. Since then, O157 has

taken part in a series of biological mergers and acquisitions that left it as vigorous as one of today's giant pharmaceutical houses. Indeed, a 2001 study showed that O157, composed of more than 5,400 genes, picks up foreign DNA at a much faster rate than do other organisms. At some point, it acquired two deadly Shiga toxin genes after being infected by a bacteriophage, a tiny virus that insinuates its DNA into the chromosome of a bacterium. In the microbial world, phages are like squatters in Amsterdam, casually taking up residence in new bacteria, perhaps as a response to environmental stresses such as ultraviolet light or toxic chemicals. Bacteriophages are also the villains behind some of the most deadly human plagues; the genes coding for the cholera toxin, for instance, were borne on a phage. So what surrounding pressures compelled the phage carrying the Shiga toxin genes to light out for a new home in *E. coli?* In experiments on mice, Tufts University researcher David Acheson may have found the answer. When Acheson gave the animals low levels of antibiotics, the phage virus wildly replicated itself, and its magnified forces were more likely to infect other bacteria. Antibiotics also spurred the phage to pour out clouds of Shiga toxin. Acheson speculates that when farmers began the practice of feeding cattle small doses of antibiotics to spur growth, beginning in the 1950s—perhaps not coincidentally, when the first reports of sporadic HUS in children came out—they may have unleashed O157. More backing for this theory comes from epidemiological evidence. *E. coli* O157:H7 is a disease of affluent, developed nations—which also happen to be the ones that feed growth-promoting antibiotics to livestock.

What worries Acheson and other scientists is that the restless phages that manufacture Shiga toxin may jump to other disease-causing bacteria. Actually, they've already proven they're disposed to do this, having set up home in about 200 other strains of *E. coli*. One of these, *E. coli* O111:H8, in 1999 caused a massive epidemic of nausea, vomiting, bloody diarrhea, and severe stomach cramps at a high school drill team camp in Texas, sickening dozens of the 750 teenage girls who attended. Though investigators never did find where the organism was hiding, they suspect it was either in the ice the girls used to soothe their parched throats during the drills or somewhere in the salad bar. Shiga toxin phages have also landed in *Enterobacter* and *Citrobacter*—other bacteria that stir up intestinal disease. To find out just how prevalent these mysterious strains of dangerous *E. coli* may be, Acheson analyzed ground beef samples from 12 supermarkets in Boston and Cincinnati. The results came as a shock. He found Shiga toxin in a quarter of the samples—toxin produced not by O157:H7, but by other kinds of *E. coli*. And this may not be the end of their roving, Acheson warns. "Suppose something like *Salmonella* developed the ability to produce Shiga toxins. That could be an extremely deadly pathogen." Not only is *Salmonella* common, but, more than *E. coli* O157, it has a talent for quickly invading the bloodstream, meaning it could speedily convey Shiga toxins throughout the body like tiny poison-tipped missiles. Even more problematic, the antibiotics normally used to treat *E. coli* O157:H7 infections may actually aggravate the illness, by kicking phages into overdrive and stepping up their production of toxins, leading to hemolytic uremic syndrome.

Along its evolutionary path, *E. coli* also became acid resistant, so impervious to a low pH environment that it can survive the incredibly sour bath in the human stomach. Grain-feeding cattle, which supplanted traditional hay feeding after World War II, may have made the bacteria more acid resilient. Because of this acid tolerance, as few as 10 organisms are enough to cause infection. Having acquired a mean set of toxin genes, acid resistance, and other virulence properties, all *E. coli* O157:H7 needed to become a truly fearsome threat was access. That it acquired by spreading in domesticated cattle and then entering the gears of modern industrial meat production, all within the past 25 years. Unfortunately, O157 may have left the door open behind it. Other strains of *E. coli*, "if tweaked in the right way" by phages and the mobile rings of DNA known as plasmids, could negotiate the same path, says Tom Whittam, a biologist at Pennsylvania State University who has studied O157 evolution.

Research is under way on vaccines that would prevent cattle from carrying O157, and on feed additives—including competing intestinal bacteria—that would eliminate the pathogenic organism in livestock.

Thoroughly cooking ground beef to a temperature of 160 degrees Fahrenheit is the proven method of killing *E. coli* O157:H7. But in the United States, the organism retains a fighting chance because of the American love affair with rare burgers, which practically guarantees that one man's meat will be another man's poison. As a restaurant menu in suburban Dallas proudly informs its customers: "The Department of Health suggests MEDIUM-WELL for any ground beef product. Our burgers are cooked MEDIUM (PINK) unless you request otherwise."

Animal Farms

The *E. coli* O157:H7 saga shows how the denizens of an animal's GI tract find their way to our own digestive systems. This brings up a delicate point, rarely discussed in polite company, but one central to the rest of this chapter. Put simply, animal and human waste is the source of most foodborne illness. And what we eat usually becomes contaminated long before it reaches us—during processing, at the slaughterhouse, or right on the farm.

Of course, that's a resonant theme in public health. The sanitary revolution of the nineteenth century—the discovery that the diseases of squalor and overcrowding could be prevented with sewage removal and clean water—was occasioned by fear of cholera, typhoid fever, and other pestilential diseases. Before this transformative event, daily life was unimaginably filthy. "Thousands of tons of midden filth filled the receptacles, scores of tons lay strewn about where the receptacles would receive no more," observed an English medical officer in Leeds in 1866. "Hundreds of people, long unable to use the privy because of the rising heap, were depositing on the floors."

Which is precisely how the animals that become our food live today. And why, at the CDC, officials in the foodborne and diarrheal disease branch long for a sanitary revolution: clean piped water and sewage disposal and treatment—for animals. Like the nineteenth century innovations that controlled typhoid fever, animal sewage would be separated from the human food and water supply, and also from the *animal* food and water supply. "It's a paradigm shift," says the CDC's Fred Angulo. "Farmers don't consider themselves food handlers."

The site of modern meat production is akin to a walled medieval city, where waste is tossed out the window, sewage runs down the street, and feed and drinking water are routinely contaminated by fecal material. Each day, a feedlot steer deposits 50 pounds of manure, as the animals crowd atop dark mountains composed of their own feces. "Animals are living in medieval conditions and we're living in the twenty-first century," says Robert Tauxe, chief of the CDC's foodborne and diarrheal diseases branch. "Consumers have to be aware that even though they bought their food from a lovely modern deli bar or salad bar, it started out in the sixteen hundreds."

The feedlot is just the start of their fetid journey. At the head of the slaughterhouse line, a "knocker" wields a pistol-like device to drive a metal bolt into a steer's head. Other workers cut the animal's throat to drain blood, and use machines to sever the animal's limbs, tear off its hide, pull out its organs. More than 300 animals may pass through the line in an hour, each carcass weighing 650 to 800 pounds. At the slaughterhouse, writes journalist Eric Schlosser, "The hides are now removed by machine; but if a hide has not been adequately cleaned first, pieces of dirt and manure may fall from it onto the meat. Stomachs and intestines are still pulled out of cattle by hand; if the job is not performed carefully, the contents of the digestive system"—i.e., waste—"may spill everywhere."

A United States Department of Agriculture study published in 2000 found that 50 percent of feedlot cattle being fattened for slaughter during the summer months carried the *E. coli* O157:H7 bacterium in their intestines—a far higher figure than previous government estimates. Another study found that about 43 percent of the skinned carcasses tested positive before being eviscerated, suggesting that microbes were being spewed within the plant.

In early July 2000, the Excel Corporation—the nation's second-largest beef processor—allowed an Associated Press reporter to visit its huge Fort Morgan, Colorado, meat packing plant. Asked about the dangers of tainted meat reaching consumers, Excel's food safety director replied: "It's like a roll of the dice or a game of Russian roulette." Two weeks later, the face of three-year-old Brianna Kriefall, of South Milwaukee, appeared on front pages across the country. She had died from eating a slice of watermelon at a Sizzler restaurant. The watermelon had been sliced in the restaurant kitchen, on the same countertop where a meat grinder was used to convert steak trimmings—*E. coli*–contaminated steak trimmings—into hamburger. The trimmings came from sirloin meat packed in heavy vacuum-sealed bags. The bags had been shipped just a few days earlier from Excel's Fort Morgan plant.

***

Chicken farming is just as noxious. But before delving into that, a word about chickens: they're not all created equal. In the agribusiness world, there are two kinds of chickens—the broilers that give us meat, and the layers that give us eggs—and they are totally separate industries governed by different practices, riddled with different problems, and even centered in different parts of the country (the top broiler states are Georgia and Arkansas, while the top egg-producing locales are Ohio and California).

First, a look at broilers. In her book *Spoiled*, journalist Nicols Fox writes that "If chicken were tap water, the supply would be cut off." Oddly enough, the government doesn't have hard numbers on *Salmonella* and *Campylobacter* contamination rates, and what they do have is hardly appealing. A 1999 study from the USDA's Agricultural Research Service, for instance, found that 7 percent of chickens sampled at slaughterhouses had *Salmonella* and 30 percent had *Campylobacter*— but, as one scientist there admitted, those numbers are probably low. For many years, researchers assumed that clearing feces, rodents, and insects from the broiler houses where the birds live out their five to nine weeks would solve the problem. But new studies suggest that the source of chicken contamination may be more deep-rooted. The 9.5 billion young broilers that Americans eat each year are actually the fourth generation in a carefully husbanded line. Scientists now believe it's the three previous generations—the "breeders"—that regularly pass down infection. When he tested birds at the top of the pyramid —the great-grandparent breeder flocks—microbiologist Nelson Cox at the USDA's Russell Research Center found that 36 percent were positive for *Salmonella*. Cox suspects that these birds transmit pathogens to subsequent generations by contaminating their own eggs with feces that carry high levels of *Salmonella, Campylobacter, Listeria*, and *Clostridium perfringens*. Because the hen's body temperature is quite warm—between 104 and 107 degrees Fahrenheit—she usually lays her egg on a day in which the air is cooler than her body. That temperature gap forces bacteria on the porous surface of the egg to get sucked into the membrane underneath, where most organisms live contentedly while the fertile egg is incubating. When the chick pecks its way out, it eats the pathogens. "The largest contributor to contamination of a broiler flock," says Cox, "is the mother hen—the feces of the parent bird." That means the human disease on our end of the food chain won't end until farmers either clean up the three generations above, or scientists figure out how to snap the links of contamination. Vaccines may not be the answer, since they are only effective against diseases that make chickens sick—and both *Salmonella* and *Campylobacter* are benign commensals, living happily in the birds' intestinal tracts without causing harm. Another possibility, slaughtering the priceless great-grandparent breeder birds, would drastically raise chicken prices.

Now on to layers. Modern houses for egg production are avian megalopolises. In 1945 the typical henhouse sheltered 500 birds; today it can contain 80,000 to 175,000, with up to 20 houses in a single operation. (As in the livestock industry, this huge scale is a result of industry consolidation; in 1996 there were approximately 900 egg operations in the United States, compared to 10,000 in 1975.) Laying flocks stay in the same house for up to a year and a half, which means that detritus builds up. "You have a lot of everything," says Richard Gast, a microbiologist at the USDA's Southeast Poultry Research Laboratory. "A lot of birds, a lot of manure, a lot of moisture, a lot of dust. Everything that walked into that house—every two- and four- and six-legged creature—is a potential vector for moving it around."

What spreads in this tenement is *Salmonella enteritidis*, or SE, the villain behind most egg-related outbreaks. SE is a versatile bug, capable of infecting birds through two different routes. One is orally, since chickens eat feces. Another route—more troubling, because scientists haven't figured out how to interfere with it—ascends through the cloaca, the cavity in birds into which empties the products of both the intestinal and reproductive organs. SE is sucked up into the bird's reproductive tract and eventually into the ovaries. From there, it gets inside eggs even before the shell is laid down—indeed, most eggs become systemically contaminated with *Salmonella enteritidis* in this way. Just where SE came from, or why it spread so suddenly in the 1980s, remains a mystery. Found in 1 of every 20,000 eggs, SE makes French toast, Hollandaise sauce, and raw cookie dough risky culinary excursions.

Animal waste and its dangerous microbes aren't confined to the farm, of course. Manure—spread through fertilizer, irrigation water, insecticide solutions, dust, even wild birds and amphibians—gets on produce too. Typical is an outbreak that took place in 1998, when patrons of a Kentucky Fried Chicken restaurant in Indianapolis became ill with *E. coli* O157:H7. Zeroing in on KFC's cole slaw, investigators discovered that some of the cabbage came from fields supplying a Texas vegetable company—and that, during a severe drought, the fields were flooded with untreated water from the Rio Grande, where cattle had waded and relieved themselves in the irrigation canals. Similarly, at Disney World in Orlando, Florida, thousands of visitors from all over the country were believed to have been infected with *Salmonella* in 1995 after drinking unpasteurized orange juice at special "character breakfasts" at the park, in which costumed Disney characters mingle with the guests. The orange juice came from a small processing plant nearby—a plant where the walls and ceiling of the processing room had cracks and holes, and where frogs congregated near the equipment. Outside the plant, investigators found *Salmonella* in a toad, in tree frogs, in soil, and on unwashed oranges. *E. coli* O157:H7 contaminates unpasteurized cider when fallen apples touch cattle or deer waste and are then mixed with other pieces of fruit. Numerous lettuce outbreaks have occurred after the heads were exposed to cattle manure. Organic foods are hardly immune to these pitfalls. In fact, microbiologists have found more bacterial contamination on organic than on conventionally grown produce, and no one is quite sure how much composting it takes to knock off pathogens in manure. That gap in knowledge has real-life consequences. In rural Maine in 1992, a woman who abided by a lacto-vegetarian diet, consisting almost exclusively of vegetables fertilized with manure from her cow and calf, developed *E. coli* O157:H7 when she failed to wash the vegetables well enough. Through improper handwashing, she passed the infection on to three neighborhood children, one of whom, a three-year-old boy, died of kidney failure.

Farm conditions create a wide-open channel down which emerging pathogens travel from food animals and produce to people, and the modern food industry has converted a two-lane country road into a 12-lane interstate. "Salmonellosis is rare in developing countries, where sanitation is poor and diarrheal diseases are endemic, but where food production and consumption are local," writes Martin Blaser, chairman of the department of medicine at New York University, in the *New England Journal of Medicine.* Blaser's dispiriting conclusion? "Salmonellosis—with the notable exception of typhoid fever—is a disease of civilization."

And outbreaks are not so much "point source" as pointillist. Changes in agriculture and food manufacture—vaster and fewer farms, slaughter plants, and processing facilities—have given pathogens a larger stage on which to strut. In this miraculous food economy of scale, when things go wrong, they go wrong in a big way. Mass-distributed items with spotty or low-level contamination are consumed by people living far from the source. This leads to a new, insidious kind of epidemic: one with low attack rates (less than 5 percent of the people who eat the contaminated food) but huge numbers of dispersed victims. Take the massive 1994 outbreak of *Salmonella enteritidis.* Usually, SE is linked to undercooked eggs or egg products. But Schwan's ice cream, made in Minnesota and delivered to homes in all 48 contiguous states, was made from premix that had been transported to the plant in tanker trailers—trailers that had previously carried unpasteurized liquid eggs. Though the insides of the tankers were supposed to be washed and sanitized after hauling eggs, drivers sometimes skipped that laborious step. Across the country, an estimated 224,000 ice cream aficionados—mostly kids—paid the price in the largest outbreak of salmonellosis ever recorded from a single food source.

The Path of Most Resistance

The antibiotics that food animals eat can also make you sick. Though the drugs are used to fatten livestock and protect them from disease, they have the paradoxical effect on humans of breeding mean, antibiotic-resistant infections. Here's how the process works: In an animal's gut, antibiotics foster the growth of bacteria such as *Salmonella, Campylobacter,* or *E. coli* that resist the antibiotics. If you eat undercooked meat from that animal, you swallow those antibiotic-resistant bacteria, and you may or may not get sick. By themselves, drug-resistant organisms in food don't necessarily trigger symptoms, because the bugs are held in check by other bacteria in the gut. But if you happen to be taking the antibiotic to which the organism is resistant—say, tetracycline—you can get very sick. That's because the drug clears out other benign bacteria in your intestines, opening the way for the very pathogen that resists the antibiotic to run rampant in your colon and sometimes beyond.

"The reason we're seeing an increase in antibiotic resistance in foodborne diseases is because of antibiotic use on the farm," says the CDC's Fred Angulo. In the United States, an estimated 70 percent of the antibiotics produced each year—nearly 25 million pounds, according to a 2001 report—goes to food animals, in low nontherapeutic doses. Farmers mix these antibiotics in animal feed for two reasons. One is to prevent disease. The other is to promote growth and boost the conversion efficiency of feed into flesh, though exactly how low-dose antibiotics accomplish this isn't clear. It may be that these drugs kill off not only disease-causing bacteria in the gut, but also the good bacteria that compete for nutrients. What's scary is the overlapping of farm and pharmacy. Of the 19 classes of antibiotics used in animals as growth promoters, seven are prescribed for people.

When foodborne pathogens turn antibiotic-resistant, they wreak all kinds of havoc. They are more virulent, and they afflict more people because it takes fewer organisms to cause infection. Patients with antibiotic-resistant infections stay in the hospital longer. The infections especially threaten children, the elderly, and people whose immune systems are weak, such as cancer or AIDS patients: all groups likely to take antibiotics. Foodborne infections that breach the intestinal tract and enter the rest of the body trigger bloodstream or nervous system infections—for which antibiotic treatment can be lifesaving. When the pathogen resists the best drugs doctors can offer, death rates climb. Resistant foodborne infections also complicate treatment for other, unrelated infections.

For many years, farmers and regulators were resolutely skeptical that antibiotics in animals could have downstream effects in people. It took a dramatic 1983 outbreak, in which 18 people in four Midwest states came down with a ferocious strain of antibiotic-resistant *Salmonella newport*, to erase the conventional wisdom. Just before becoming ill, most of the patients happened to have taken a form of penicillin for garden-variety infections: bronchitis, earaches, strep throat. So dramatic was the link between taking the antibiotic and coming down with salmonellosis—patients were 51 times more likely than those in the control group to have taken the drug—that public health officials first suspected the antibiotic itself was contaminated. The truth was much more devious. When investigators gathered patients' food histories, they found that all had eaten ground beef shortly before falling ill. In each case, the hamburger meat had come from a South Dakota farm where beef cattle were fed "subtherapeutic" doses of antibiotics. On the adjacent farm, a dairy calf had died of *Salmonella newport.* Investigators conjectured that the dairy herd somehow transmitted the bacterium to the beef cattle which, being fed small doses of tetracycline antibiotics, went on to develop resistant strains of the organism. The infected beef cattle contaminated at least 40,000 pounds of ground beef with antibiotic-resistant *Salmonella*; without a doubt, many more people ate it than came to health officials' attention.

Today, the link between antibiotic use on farms and human disease is richly documented. In Muslim countries, resistant foodborne bacteria in people are almost always identical to microbes found in poultry—not surprising, since pork is banned. In the Netherlands, vancomycin-resistant enterococci, or VRE—a looming problem in hospitals—are noticeably absent from the intestines of strict vegetarians. Around the world, governments have been embroiled in the question of whether food animals should get the same antibiotics prescribed for people. This question is bound to loom larger, since—as you will learn in the next chapter—many public health officials find antibiotic-resistant infections to be the most terrifying prospect on the horizon.

Mad Cows and Englishmen

Another insidious foodborne infection may be looming on the U.S. horizon. New variant Creuzfeldt-Jakob disease—the human form of mad cow disease—has proceeded narrowly and stealthily through a food chain whose links are masked by intensive food production and globalization. At this writing, more than 100 cases, invariably fatal, have been reported, mostly in Britain with a handful in Europe. What researchers don't know is whether these represent the waning aftermath of a narrowly averted public health disaster or the first rumbles of a terrible storm.

"Mad cow disease" is a term that had not even been coined in late 1984, when a veterinarian called to a farm in West Sussex, in southern England, found a dairy cow displaying "a variety of unusual clinical manifestations": panic, aggression, a staggering gait. By the spring of 1985, more cows came down with the mysterious malady, later named bovine spongiform encephalopathy, or BSE, the giveaway marker of which was brain tissue that resembled Swiss cheese. By the late 1980s, scientists began piecing together the puzzle. The epidemic likely began as a foodborne outbreak among livestock. In the early 1980s, cattle were fed remnants from sheep infected with scrapie, a brain-wasting disease (discarded animal parts are considered a cheap source of protein that increases milk production). In a further perversion of nature, the inedible parts of these infected cattle were themselves made into meat and bone meal for other cattle, a thrifty practice that not only permitted the recycled infectious agent to amplify but also to adapt to its new host. Changes in the rendering process also helped the agent survive. Before 1981, the carcasses of ruminants had been subject to high heat and organic solvents to remove fat and disarm disease-causing proteins, such as viruses. But that year, various economic factors persuaded manufacturers to turn down the heat and cut out the solvents, allowing the yet-to-be-discovered infectious agent to escape inactivation. Though hundreds of thousand of cows would eventually be diagnosed with BSE, and millions of animals destroyed as a precaution, the British government was stalwartly optimistic that the epidemic would stay put on the farm. As the Southwood Report noted in 1989, "It is most unlikely that BSE will have any implications for human health." By 1996, after a new variant of the neurological affliction Creutzfeldt-Jakob disease—featuring bizarre behavioral and personality changes, staggering, and dementia—appeared in startlingly young patients, the scientists changed their tune, to the horror of a carnivorous nation. "Beef is one of the great unifying symbols of our culture," lamented a *Guardian* editorial. "The Roast Beef of Old England is a fetish, a household god, which has suddenly been revealed as a Trojan horse for our destruction."

How much beef contaminated by prions—abnormally folded proteins—do humans have to consume to become infected? No one knows. And no one knows how long the incubation period is for new variant Creutzfeldt-Jakob disease, though judging from events in Britain it seems to be at least 10 to 15 years. No one knows how much BSE-infected beef was slaughtered for human consumption before the epidemiologic puzzle was pieced together—perhaps 750,000 animals, perhaps a million; the UN estimated that at the height of the mad cow epidemic, Britain dumped 500,000 tons of untrackable bovine byproducts on Western Europe and other nations. No one knows whether mad cow prions have infected people through blood or organ donations, contaminated surgical instruments, or consumer products and drugs that contain bovine material. The upshot of all this uncertainty is that no one knows where we are on the epidemiologic curve: at the end or the beginning of an outbreak? So murky is the science, Oxford University's esteemed epidemiologist Roy Anderson calculated that human cases could conceivably range between 63 and 136,000, while a British government study put the high-end figure at 250,000.

Will the human variant of mad cow disease turn up in the United States? "The odds are that sooner or later we will see a case here," says CDC director Jeffrey Koplan, "whether it's an imported one, whether it's home grown, whatever. None of us should be surprised if we have a case in the next week or the next ten years." As of the fall of 2001, no cases have been reported. To prevent the spread of BSE to American farms, the U.S. government in 1989 banned the importation of live cows and sheep. Since then, it has erected regulatory fences to screen out other bovine products and has upgraded surveillance for brain-riddling spongiform diseases in domestic animals and humans. The American Red Cross has tightened its blood donation rules for people who have been to Europe. Critics say these measures aren't enough— that, to borrow from W. C. Fields, the United States has failed to "take the bull by the tail and face the situation." The U.S. still imports biomedical products, for instance, that contain materials made from ruminants in countries harboring mad cow disease. In 2001, the Food and Drug Administration reported that companies were using ingredients from BSE countries to make nine widely used vaccines, including those for polio, diphtheria, and tetanus. The FDA has also failed to regulate dietary supplements such as those claimed to stimulate energy, sexual vitality, and memory—all of which can contain nervous system, organ, and glandular tissue from cattle. And the surveillance net for potentially infected cattle in the United States has big holes, while inspection of feed manufacturers and rendering companies is lax. Meanwhile, some scientists worry that spongiform disease could strike Americans, not through the consumption of beef, but of hunted wild animals such as deer and elk, which are succumbing to another prion-related epidemic, chronic wasting disease.

In echoes of the mad cow crisis, another agricultural infection struck Britain in 2001—foot and mouth disease, one of the most contagious of all animal diseases. How did the culpable agent—in the same family as the common cold virus—get there? Likely from contaminated meat smuggled into Britain from countries where the disease is rife. The British army admitted supplying untreated waste food to a pig farm in Northumberland, where the virus incubated and then wafted over air currents to a flock of sheep. By the time the disease was identified days later, the virus had spread all over the country through markets and dealers. The government response was swift and shocking. Bonfires of livestock carcasses shot flames into the night sky—one writer described the giant pyres as "archaic precautions." Europe and the United States, long protected against the infection, went on red alert, disinfecting the shoes of hundreds of thousands of arriving airline passengers from the British Isles—another reminder that the world is not just a global village, but a global pathosphere.

Health Foods and Stealth Bacteria

Consider the alfalfa sprout: from a moist mat of tendrils, a sinuous pale stem rises to a dark green, double-bladed capital. Sprout growers proudly call their crop a "living food," in contrast to the butchered, disease-bearing livestock that are a staple of the American diet. They boast that one ounce of sprouts contains more protein than a pound of steak and is loaded with vitamin C. Prefiguring the counterculture, Captain James Cook raised lentil sprouts to stave off scurvy, until colonial economics made limes more favored on the high seas.

Food microbiologists hold a somewhat less romantic view of this '60s holdover. "Sprouts are about as hazardous a food as you can get," says Mike Doyle, director of the Center for Food Safety and Quality Enhancement at the University of Georgia. "You've got water, you've got the right temperature, and you've got nutrients. Those are the three things you need for growing harmful microorganisms." From a public health perspective, sprouts perfectly illustrate what can go awry when a "living" food that happens to be contaminated multiplies and goes forth into the food supply. And it shows that whenever a new food technology arrives on the scene, its impact on foodborne pathogens must be thoroughly pondered.

Between 1970 and 1999, the sprout industry grew at an annual clip of 10 percent. Many of its entrepreneurs were devoted vegetarians and organic gardeners who eschewed chemical additives such as pesticides, hormones, or disinfectants. By the end of the century, 15 different varieties of sprouts appeared in U.S. stores, from alfalfa and clover to more exotic species such as sunflower, wheatgrass, and anise. What these highly independent small businesspeople overlooked was that *Salmonella* and other pathogens are often present at the creation— lurking in the crevices of sprout seeds.

Described by Nicols Fox as "the cockroach of the microbial world," the *Salmonella* genus is hardy, tenacious, adaptive, omnipresent, and virtually impossible to eradicate. *Salmonella's* more than 2,400 serotypes can survive an amazing variety of conditions, from fish meal to fresh orange juice, from a hen's ovaries to a turtle's toes. *Salmonella typhi* causes typhoid fever, spread by contaminated water and food. Other *Salmonella* serotypes were named for the places where they were isolated, little flags on a global map of bad meals: *S. newport, S. muenchen, S. heidelberg,* and so on. Given its cosmopolitan ways, it's no surprise that *Salmonella* turned up in sprout seeds. These seeds can be contaminated in the field from dirty water, runoff from nearby farms, and animal fertilizers, and from the feces of birds and rodents anywhere along the line from growth to storage to shipment. *Salmonella* organisms can survive for months under the dry conditions in which seeds are stored. The presence of even a few salmonellae on seeds is dangerous, because sprouting's moist, warm growing conditions encourage bacteria to surge. In three to five days, ten organisms on a single seed can grow to more than 220 million in a typical eight-ounce container. You wouldn't be able to see them with the naked eye or taste them in a salad—or wash them off, since they are found in the tissue of the sprout, not just on the surface. And while the FDA recommends that sprout growers first rinse seeds with chlorinated water to bring down bacterial counts, the method doesn't guarantee that pathogens will be eliminated before the crucial sprouting step. Making matters worse, sprouts are rarely cooked—a traditional kill step for pathogens—before being eaten.

The tipoff to the hazards of sprouts came in 1973, when a home sprouting kit spread *Bacillus cereus*, a toxin-forming organism found in soil. Since then, there's been a steady stream of outbreaks, mostly from clover and alfalfa sprouts. In 1996 in Japan, white (daikon) radish sprouts infected an astounding 10,000 people with *E. coli* O157:H7. Just how rife *Salmonella* is became apparent in 1999, when the National Center for Food Safety Technology, a lab affiliated with the FDA, tested a new *Salmonella* rapid detection assay. Researchers inoculated *Salmonella* in some seeds and kept the control group of seeds clean. Perplexingly, the rinse water for both groups of seeds turned up positive for *Salmonella*. At first, the scientists thought the new test was giving false positive results—until they realized that the supposedly "clean" seeds in the control group not only were contaminated with *Salmonella*, but came from the same lot that was causing an outbreak in the Midwest.

Not until the spring of 1995 did U.S. public health officials wise up to the true dangers of sprouts. That year, the CDC kicked off a new method of discerning whether *Salmonella* cases were rising in this country. "SODA," or *Salmonella* Outbreak Detection Algorithm, compared new state health department reports of *Salmonella* with the number that would be expected, based on historical data. The very next month the new system fulfilled its promise, revealing an international epidemic of *Salmonella stanley* that had been brewing for months, an outbreak that reached from Arizona and Michigan to Finland and probably Canada as well. Finnish officials who had already investigated the outbreak found that most of the patients remembered eating alfalfa sprouts before becoming ill. Tracing back the sprouts to their seed origins yielded a typically Byzantine course of events behind a modern international outbreak. The sprouts eaten by the American patients came from at least nine different domestic growers, all of which had bought seeds from a single supplier to the U.S. market. This supplier had bought its seeds from various sources, including a shipper in the Netherlands. The Dutch shipper had mixed several lots of seeds purchased from Italy, Hungary, and Pakistan—and through an intermediate supplier, shipped some of these mixed lots to Finland. Ultimately, investigators estimated, between 5,000 and 24,000 people became sick in this low-profile, globe-spanning outbreak.

Thinking Globally

Parsley is gharsley.

— Ogden Nash, *Further Reflections on Parsley*

As the sprout outbreak showed, national borders dissolve in the face of foodborne pathogens. The General Accounting Office estimated that in 1995, one-third of all fresh fruit consumed in the United States was imported, a testament to lower production costs abroad and rising consumer demand for year-round fruits and vegetables. Seasonally, more than 75 percent of fresh produce may be harvested beyond U.S. borders. Though most imported food is wholesome, it is inevitable that, as the global food market grows, so will the number of international outbreaks. The question is: whose food standards will reign, those of America or those of our aspiring trade partners?

In the summer of 1998, when Minnesota health officials were barraged with outbreaks of gastrointestinal illness, they weren't thinking about these questions. *Shigella sonnei*, a diarrhea-causing bacterium, had struck hundreds of patrons from two Minneapolis restaurants that apparently had nothing in common. One served Up North cuisine in Bunyanesque portions, while the other was part of a suburban horse and hunt club. The timing couldn't have been worse. The health department was already besieged with a mysterious epidemic of enterotoxigenic *E. coli*, or ETEC—commonly known as "traveler's diarrhea"—that had sickened more than 50 at a Twin Cities restaurant and a lake resort. It was strange, this gastrointestinal tempest. In a state reputed to have the most vigilant and progressive public health department in the country, where national outbreaks were solved before other states even had an inkling something was amiss, health officials were stumped.

The *Shigella* bacterium is so potent that its misery can be transferred in a handshake. As few as ten organisms, a remarkably low infectious dose, can cause infection. Bacteria invade the cells lining the colon, inflame the intestines, and eventually bring on fever, doubling-over abdominal pain, nausea, vomiting, and diarrhea. In the mid-1990s, a strain of *Shigella sonnei*, the predominant species in the United States, spread through eight communities of traditionally observant Jews in North America, fanning out in informal day care, ritual handwashing, and steam baths—more than 1,000 cases in a closed but global neighborhood. Today, the organism is most notorious for racing through day care centers, where toddlers and young children, whose personal hygiene practices are less than rigorous, spread infection—often to their diaper-changing parents and other adult handlers, who in turn infect others.

In Minnesota's 1998 outbreak, suspecting at first that restaurant workers were the source of infection, investigators embarked on their standard gumshoe procedure. They interviewed the sick patrons and their unaffected dinner companions, hoping to pinpoint what the shigellosis victims had eaten. No food seemed resoundingly guilty. More surprising were the results from the state lab's genetic analysis of the bacterium found in patients' stools: the patrons of both restaurants had the same strain of *Shigella*. And this was a brand-new strain, not one that had been circulating around town. That meant the disease wasn't spread by a few infected workers, usually the first guess in cases like these, but rather by a common food. And this food, whatever it was, must have come from the same farm. The investigators recrunched their data. They broke down every menu item at each restaurant into its constituent ingredients, trying to find what single element in two different kitchen larders could have been the culprit. The item that finally jumped out of a mess of statistics was a stunner: parsley. Specifically, chopped fresh curly parsley, decorative rather than delectable, the culinary equivalent of wallpaper.

Minnesota health officials launched a national investigation, dispatching a picture of the *Shigella*'s DNA on PulseNet, an electronic server that the CDC and state public health labs use to compare the genetic profiles of foodborne bacteria sampled at different sites. PulseNet is a public health version of the FBI's 10 Most Wanted List, except that it features thousands of microbial criminals, and the mug shots it distributes are not the haggard faces of human reprobates but a column of shadowy bands representing unique stretches of DNA. Minnesota's PulseNet image was accompanied by a question: Were other states also apprehending this strain of *Shigella sonnei?*

In Los Angeles, alarms went off. The city had seen two perplexing outbreaks of *Shigella sonnei,* one at an upscale steak-and-pasta place, the other at a Middle Eastern restaurant. Prompted by Minnesota's message, investigators returned to the evidence and found that the strains from their two outbreaks not only matched each other, but also matched Minnesota's—and that sure enough, the sick customers had all eaten parsley, either as a garnish or mixed into tabouli and falafel.

Soon Massachusetts chimed in with *Shigella* isolates that matched Minnesota's. There had been an unsolved *Shigella sonnei* outbreak at a restaurant-and-bar joint north of Boston. The sick customers had all eaten grilled chicken breasts—which, investigators now learned, were liberally sprinkled with minced parsley.

A Marathon Key, Florida, restaurant was ground zero for *Shigella sonnei*. When investigators interviewed patients and other customers, they found that all had eaten dishes with fresh parsley—in fact, it was nearly impossible to find anyone who had *not* eaten parsley. The decorative sprigs were scattered over every entrée as if they had talismanic powers: on the baked mango snapper and fried grouper fingers, on the shrimp primavera and veal piccata, on the chicken cordon bleu and steak au poivre.

Canadian authorities also phoned: two outbreaks of *Shigella sonnei*, one in Alberta and one in Ontario, with the same strain as the one bombarding the States.

All told, 486 confirmed cases of shigellosis—and that was merely the count of diagnosed cases. In Minnesota alone, authorities estimated the disease actually struck at least 600 people, including visitors to the Goodhue County Fair. As it happened, several victims of the restaurant outbreaks, still suffering symptoms, had visited the fair and used the restrooms. The antiquated septic system leaked into the fair's drinking water, extending a parsleyborne *Shigella* outbreak into a secondary waterborne epidemic (much like the events behind the massive 1999 *E. coli* O157:H7 outbreak at an upstate New York county fair that killed two and sickened more than a thousand).

So how had the parsley become contaminated in the first place? By Labor Day, Minnesota health officials had followed the product's paper trail to a farm in Mexico. When state epidemiologist Mike Osterholm alerted the FDA, he was rebuffed. "I would have gotten more response out of talking to a rock. They said, 'You don't have absolute proof it's there.' I said, 'I know we don't have proof, in the sense of absolute. But our traceback takes it there.'" Not until that October did the CDC and FDA conduct an official "traceback," a tedious and time-consuming physical examination of records from purchasers and assorted middlemen. The trail for seven of the eight outbreaks led to a vegetable growing operation south of Ensenada, Mexico, on the Baja Peninsula, about 100 miles south of the California border. Sprawled over 1,000 hilly acres hugging the Pacific, drenched in sunlight year-round, the farm grew tomatoes, celery, cilantro, green onions, radishes, and lettuce—a salad bowl inventory that would later add a twist to the story. Also under cultivation were about 55 acres of parsley—from a distance, brilliant emerald squares amid the brown hills. The herb was harvested early in the morning, chilled and packed in ice, and dispatched in a refrigerated tractor-trailer by 2 p.m.

During any traceback of contaminated produce, investigators are keenly curious about water. Water is used to irrigate crops, wash them, chill them, ice them. And water can come from many places—rivers, streams, irrigation ditches, wells, sewage treatment facilities, runoff, industrial washes, and ice. Water containing animal waste can inadvertently wash onto fields from swollen rivers or be mixed with pesticides. People suffering gastrointestinal illness can contaminate water or ice just by dipping their not-quite-clean fingers in it. When a CDC environmental engineer went to the Mexican farm, he checked out all these possibilities. He examined the irrigation system, fed by groundwater wells sunk in an old riverbed. He looked at the portable toilets in the fields, and found no toilet paper inside—to curb theft, supervisors required migrant workers to ask for it before they went to the bathroom. Yet while these conditions were potential public health nightmares, they were not what bothered him most.

What bothered him was the packing shed, where the freshly picked parsley, radiating heat after being packed in the field in wax boxes, was chilled and sent out. First, workers loaded the boxes on a conveyor belt. The belt traveled through a hydrocooler, a kind of Jacuzzi for vegetables, the job of which was not to clean the parsley but to spray on chilled, recirculated water to keep the produce crisp and fresh-looking and prolong its shelf life. The water came from the municipal system of a nearby village. After the parsley was chilled with the water spray, workers shoveled ice on top from a portable ice-making machine before closing the box, then dumped more ice on each pallet before loading the trucks. The ice was also made from the town water supply.

Later, the CDC team would learn that the farm's migrant workers, though poor and uneducated, and housed in long narrow barracks surrounded by barbed wire, nevertheless demanded and received bottled drinking water. Soon it became clear why. Vandals frequently disabled the chlorinator for the town water supply, located in a village higher up the mountain. Several times that summer, the chlorinator had been out of service. Usually authorities were able to fix the problem quickly, but a month or so before the investigators arrived, shotgun blasts had damaged the facility beyond repair, and it had stayed that way. The hydrocooler thus sprayed untreated town water onto the produce and recirculated what was left. And though workers say they added chlorine, it probably didn't help—the dirt from just one box of parsley could have absorbed all the chlorine that otherwise would have disinfected an entire lot. And that's probably what happened. The water in the hydrocooler was contaminated either in town or by the newly picked parsley itself. Either way, effluvia from one tainted box could have dripped down on scores of other boxes, spreading *Shigella* across the lot. Once the boxes were trucked away, *Shigella* could—and apparently did—spread thousands of miles.

Minnesota officials following the paper trail found that their concurrent epidemic of ETEC traveler's diarrhea came from parsley on the same farm. And that September, in northern California, more than 300 people suffered severe diarrhea and abdominal pain after eating at a Mexican restaurant. DNA fingerprinting turned up the very same *Shigella sonnei* from the very same farm—but this time, in the cilantro used in a salsa dip. The hydrocooler was bad news for every green thing that passed through it.

Unlike most outbreaks, all the pieces fit neatly, and yet there was one nagging question: Why didn't *whole* parsley cause outbreaks of shigellosis? At the University of Georgia's Center for Food Safety and Quality Enhancement, scientists tried to find out. They bought a few bunches of raw parsley from a local supermarket, inoculated the leaves with *Shigella sonnei*, and subjected the herb to a range of preparation methods and temperatures to observe how the bacterium behaved under different conditions. It turned out that mincing the parsley created the perfect medium for the bacteria to grow, because the fluid released from broken tissue cells was loaded with nutrients. While *Shigella* organisms diminished in the refrigerator, they multiplied with abandon when chopped parsley remained at room temperature. And that's just what happened at most of the outbreak restaurants: kitchen workers chopped up a large batch of parsley in the morning and let it sit out in a big bowl during the day, grabbing a handful to garnish each dish. With the best of culinary intentions, they had cultivated little *Shigella* farms.

The parsley outbreak was only one of many warning shots. Increasingly, we are exposed to fruits and vegetables potentially grown in polluted waters or in countries where food safety regulations and working conditions are less stringent than our own. This was the lesson in 1997, when nearly 300 schoolchildren and teachers in Michigan and Maine were infected with hepatitis A—an often chronic debilitating virus that attacks the liver—after eating frozen strawberries at school. All the fruit had come from a Mexican farm on which strawberry pickers had only a few latrines. The only places they could wash their hands were on trucks that circulated through the fields. With bare hands, the pickers severed the strawberry stems with their fingernails. As the CDC's Rob Tauxe told Nicols Fox, "Those are the hands that feed you. And it might actually matter whether they are washed or not or whether they have a latrine. If we are interested in the safety of our food, then we have to be interested in the living conditions of the people who handle it."

We live in a world economy, and there's no going back. Yet imported produce brings us exotic microbes to which we lack natural immunity. Currently, FDA inspectors examine less than 2 percent of food coming into the United States. And though the agency can seize imports at the border, it can't bar imported food just because it comes from a country with questionable food safety rules. In fact, in the wake of recent trade agreements, there's political pressure not to bar such imports, since the action could be interpreted as U.S. protectionism— "nontariff barriers" to free trade—masquerading as food safety policy. Food safety advocates have called for uniform microbial standards for all produce, imported and domestic. It's a recognition that, as Americans eat more and more from a global plate, the question of whether we have "the world's safest food supply" is becoming meaningless. We *have* the world's food supply.

The Microbial Underworld
(and the Difficulty of Getting Convictions)

Just as DNA fingerprinting has supplied crucial evidence to seal criminal convictions, it has revealed surprising connections in a teeming and, until quite recently, unsuspected microbial underworld. Without this technology, the international outbreak of *Shigella* in parsley—

a perishable produce that quickly passes through the food distribution system—would never have been found. If it hadn't been so meticulously traced, the hydrocooler in Mexico may have continued to contaminate fresh fruits and vegetables. Public health officials would have been in the dark about this unusual vehicle for transmitting disease, and might not think to ask about it in future outbreaks. More broadly, insights into the fundamental biology of foodborne infection would have been lost. Outbreaks, after all, are public health goldmines. Without recognized clusters of cases and a warm trail of clues, investigators usually can't find the source of a foodborne outbreak—or ways to prevent the next.

Each day, scores of DNA fingerprints from all over the country fly over PulseNet to a network of public health laboratories dedicated to uncovering foodborne outbreaks. Using a method called pulsed-field gel electrophoresis (PFGE), the labs transform bacterial isolates into a vertical arrangement of bands, from charcoal black to faint gray, representing known sequences of DNA—a kind of genetic bar code. The technique is well suited to investigations of foodborne bacteria because, even after many generations spreading and dividing in both animal and human hosts, bacteria descended from the same parent still contain virtually the same genetic material. Lookalike PFGE patterns in two different samples suggest that the bacteria came from a common source, such as a widely distributed contaminated food. The fingerprints are entered into an electronic database at state or local health departments, and are also transmitted to the CDC, where they enter a closely monitored computer. Currently, PulseNet labs can fingerprint *E. coli* O157:H7 as well as other forms of *E. coli, Shigella sonnei,* nontyphoidal *Salmonella, Listeria monocytogenes, Campylobacter jejuni,* and others. But in the next few years, gene chip technologies will truly open the frontier, enabling labs to diagnose dozens of bacterial, viral, and parasitic pathogens in a single test.

Because it reveals camouflaged epidemics triggered by widely distributed products with spotty and low-level contamination, PulseNet is perfectly suited to the foodborne outbreaks of the future. Until recently, the scattered cases of illness caused by these subtle outbreaks were deemed "sporadic": separate and random, with no known link. PulseNet connects the dots. And because of PulseNet's sensitivity, investigators are finding smaller and smaller clusters of cases, and more of them—so many, in fact, that CDC epidemiologists are, like hurricane-watchers, giving the outbreaks names. This doesn't necessarily mean disease is increasing, but that the ability to detect it is improving. Between 1995 and 1999, for instance, the median size of a detectable *E. coli* O157:H7 outbreak dropped from 27 cases to 9. From a public health standpoint, that's good, since the more clusters officials can pick up, the more chances they have to follow leads back to the cause.

PulseNet's leads have also startled public health sleuths. What before may have looked like a single outbreak sometimes proves to be composed of several separate and concurrent outbreaks. And what used to seem like unlinked sporadic cases now sometimes turn out to be part of a dispersed epidemic with a single cause. "We had this old belief that if we were good public health people, we could get in there when the fire started and we could save the town," says Mike Osterholm. But with the old technologies, health officials noticed only the raging fires wiping out city blocks. Today, with DNA fingerprinting, they can see all the little house fires—too many fires, in fact, to put out. State and local health departments, traditionally short-staffed and underfunded, can't possibly put out or even investigate every conflagration.

PulseNet has also spurred an almost philosophical shift in public health. In a postmodern turn of phrase, the CDC's Eric Mintz refers to a new landscape of "complex meta-outbreaks," in which many more foodborne illnesses are tied to sources of contamination far back in time and space. DNA fingerprinting may reveal a dense web of microbial connections between farm animals and produce operations and food processors and the tens of millions of Americans each year who get sick from what they ate. After all, foodborne illness doesn't happen by accident. "I used to hold that there was no such thing as a sporadic case of an infectious disease; there were only patients who were not part of a recognized outbreak," Mintz says. Now, he'd extend this line of thinking even further. "Are there really *any* 'sporadic' outbreaks?

Or are all outbreaks related to other meta-outbreaks that we are now more able to discern?"

Even if investigators were able to tie together individual cases and outbreaks in a grand theory of everything, they wouldn't necessarily be able to act on that knowledge. Few investigations end as neatly or as gratifyingly as the case of the *Shigella*-ridden parsley. Sometimes DNA fingerprinting raises more questions than it answers, or presents bureaucracies with challenges that they either won't or can't face. Unlike in criminal investigations, DNA fingerprints aren't enough to close a case of foodborne illness, since there is a tiny chance that identical bacterial fingerprints may come from unrelated sources. Epidemiologists must also prove that the patients all ate a particular food and, for liability reasons, federal regulators must prove that the product was indeed contaminated.

The sometimes tricky business of closing a PulseNet case was underscored during an outbreak in the United States in late 1998. That November, the CDC received calls from several states reporting an increase in *Listeria monocytogenes* cases. Named for the British surgeon Joseph Lister, the father of modern hospital antisepsis, *Listeria*—a small bacterium that tends to line up side by side, like a regiment of recruits—is almost impossible to eradicate once it makes its home in a meat processing plant. Ubiquitous in the environment, it is found in soil, water, sewage, decaying vegetation, and many species of birds and mammals. Reports of *Listeria* are usually the kickoff for frustrating and ultimately fruitless investigations because the organism has a long incubation period; the lag time between eating a contaminated food and developing symptoms can be one to two months. That makes gathering food histories difficult, since most people can't remember what they ate yesterday, let alone eight weeks ago. More perplexing, the symptoms of listeriosis don't always suggest foodborne infection. In fact, not until 1981—after an outbreak in Canada killed 41 percent of those infected, and 34 women suffered stillbirths or miscarriages or gave birth to ill infants—did researchers even realize *Listeria* was transmitted by food. The victims had eaten commercially prepared cole slaw made from cabbage grown on a farm where listeriosis had killed sheep. If a mother's infection spreads to the fetus through the placenta, *Listeria* can cause meningitis in infants and newborns. *Listeria* infection in the nervous system causes a severe headache, stiff neck, loss of balance, and convulsions. All told, *Listeria monocytogenes* kills 1 in 5 people it infects, making it one of the most deadly foodborne agents known. Each year in the United States, more than 2,500 persons fall ill and more than 500 die. But while between 2 and 7 percent of processed meats carry *Listeria*, some strains may be less virulent than others, which is why there aren't constant *Listeria* outbreaks. "Chances are most Americans eat *Listeria monocytogenes* at least once a week," says the CDC's Paul Mead, "and yet they don't get sick that often."

The 1998 case was one of the deadly strains. By late November 1998, the death toll was mounting. Investigators had no clue where the disease was coming from. Questionnaires eventually ferreted out what the victims had in common: hot dogs. But which brand? Many had been mentioned. Here, a lucky break—also the legacy of DNA technology—pointed investigators toward the answer. That fall, Cornell University happened to have taken genetic fingerprints from pieces of smoked chicken that had sent several members of a New York family to the hospital. The DNA fingerprint perfectly matched the hot dog outbreak strain—and both, it turned out, came from Bil Mar Foods, a division of the Sara Lee Corporation, the biggest purveyor of packaged meats in the country.

By mid-December, 15 people scattered across the country had died in the outbreak and scores were ill—though with the slow trickle of reports from state health departments, the CDC wouldn't be able to add up the numbers for weeks. Ultimately, the agency calculated that the outbreak killed at least 21—15 adults and 6 stillbirths or miscarriages—and made more than 100 sick in 22 states. But the CDC and USDA tussled over whether there was enough evidence to implicate the company—while CDC insisted there was, USDA officials fretted about the legal consequences of wrongly accusing Sara Lee. Understandably confused by the government's mixed signals, Sara Lee officials called Mike Osterholm, then the Minnesota state epidemiologist, who had been at the helm of many high-profile foodborne illness investigations. Osterholm was convinced the company should carry out a recall. As he explained to the *Washington Post,* in a typical metaphorical flourish, "I said to Sara Lee that this was like driving down a highway and seeing a herd of deer. You can put on the brakes now and you'll still hit them. Or you can hit them at full speed and the deer will come through your windshield."

On December 22, Sara Lee recalled 15 million pounds of hot dogs and deli meats. But the USDA, contrary to its own policies, did not deliver a timely warning to the public; instead, it allowed Bil Mar to issue a weakly worded press release that did not mention the serious nature of the illness, or the deaths. A few days later, the CDC team isolated the *Listeria* strain found in a package of unopened hot dogs in the refrigerator of one of the Michigan patients—a woman who had given birth to a child with sepsis. The strain matched the bacterium that was quietly killing Americans.

But the government's earlier timidity had deadly consequences. In the midst of the holiday season and as presidential impeachment dominated the news, Americans never really took note of the outbreak. In Columbus, Ohio, Lisa Lee and her fiancé, unaware of the recall, continued eating Sara Lee deli meats. Lee was in the fifth month of her first pregnancy, expecting twins. That January, feeling feverish and wretched, she went to an emergency room, where a doctor diagnosed her with flu and sent her home. A few days later, she returned with a high fever. After unsuccessfully trying to stop her contractions and prevent miscarriage, doctors induced a long labor. Lee gave birth to twins, a boy and girl, both stillborn. "If we knew it as soon as Sara Lee knew it," she told an interviewer, "we might have had a chance."

As it happened, employees at the Bil Mar plant *did* know about the contamination, long before the recall. The plant had had trouble with condensation in the hot dog production room. When franks are cooked to a high temperature and then chilled, steam rises to cooling elements in the ceiling. Those cold, moist conditions make a perfect home for *Listeria*—a psychrophilic, or cold-loving, bug—which thrives on walls and ceilings and can survive for years on biofilms and metal surfaces. When the steam condenses, it rains down on whatever happens to be sitting on the production line, washing *Listeria* along with it. On the July 4th weekend, Bil Mar workers had ripped out cooling units from the ceiling and, because the units were too big to push out the door, cut them up with a chain saw before forklifting them away. Immediately afterwards, routine bacterial samples came back positive. The construction probably raised dustborne *Listeria* that drifted through the factory to the deli meat section, under the same roof but three acres away. From the plant, *Listeria* was dispatched across the country. In response to this dramatic foodborne epidemic—the most lethal in the United States in 15 years—the USDA in 2000 took steps to require *Listeria* testing in processing plants. In 2001, Sara Lee pleaded guilty to a misdemeanor charge that it had produced and distributed tainted meat. It agreed to pay $4.4 million to settle civil and criminal charges.

Food Fights

The very things that make America's food system extraordinary—its mammoth production levels, its stunning variety of imports, and its low cost—are what compromise safety. Mass production spreads risk, imports bring global pathogens to our doorstep, and historically low cost makes farmers and regulators reluctant to pursue safety measures that could raise prices. It's not the ideal way to cultivate the world's safest food supply—a claim that food safety advocates say is overblown anyway. Denmark, Sweden, and the Netherlands, for instance, have taken far more vigorous steps to eliminate pathogens.

Even if America's were the safest, it's not safe enough. So whose fault is that? Throw together all the parties involved in food safety—scientists, company officials, regulators, advocacy groups, victims and their families—and the discussion inevitably boils down to this: How much responsibility for preventing foodborne illness belongs to the individual and how much to government and industry? Those who lean toward individual responsibility point out that it's often mistakes close to home—either yours or someone else's—that inflict the

damage. Most cases of *Campylobacter*, for
instance, simply come from undercooking
poultry or letting the raw bird or its
juice touch another food. Most cases
of *E. coli* O157:H7 come from
eating ground beef heated only
to pink or pallid gray. "Raw
meats are not idiot-proof,"
says USDA microbiologist
Nelson Cox. "They can be
mishandled and when they
are, it's like handling a hand
grenade. If you pull the pin,
somebody's going to get hurt."
But while some might question

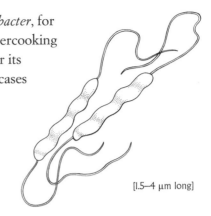

[1.5–4 μm long]

Campylobacter jejuni
Agent of acute gastroenteritis

the wisdom of selling hand grenades in supermarkets, Cox believes
that blame grows less actionable the farther back you go in the food
chain. "Who are you going to sue?" he asks. "Are you going to sue
Kroger? Or are you going to sue the man who transported it? Are you
going to sue the man who processed it? Or the one who grew it? Or
had the breeder flocks? I think the consumer has the most responsibility but refuses to accept it."

Admittedly, food abuse is rampant. Surveys show that more than
one-quarter of respondents don't wash cutting boards after cutting
raw meat, nearly a quarter prefer their hamburgers pink, and half eat
raw or undercooked eggs. The faster we forget the lessons of an agrar-
ian culture, and the more we rely on packaged and microwavable food,
the more we are apt to mishandle raw ingredients in our own kitchens.
Moreover, changes in work and family life have led us to cook less and
eat out more—Americans spend half of their food dollar away from
home, and nearly 14 percent on fast food—exposing us to the minis-
trations of young and often even more naive food handlers. A 1999
study from Los Angeles found that large restaurants—those with more
than 400 seats—were more than seven times as likely to have a com-
plaint lodged against them as a restaurant with fewer than ten seats. It

wasn't just because the restaurants had more customers, but because they had varied menus, creating more chances for cross-contamination behind those swinging kitchen doors.

Inexperienced kitchen help who may not receive training in sanitation or hygiene are also perfectly positioned to broadcast their own foodborne ailments. Because wages in these jobs are low, and insurance and sick leave nonexistent, "people are encouraged to work while they're sick," says Kirk Smith, who supervises foodborne disease investigations at the Minnesota Department of Health. In addition, many food workers emigrate from countries where intestinal infections are endemic. For all these reasons, employees suffer unusually high levels of enteric disease. While up to 0.5 percent of individuals in the general population may carry an intestinal parasite, in restaurants that have triggered foodborne outbreaks, up to 18 percent of food handlers have been shown to suffer intestinal infections. The public health literature is replete with reports of kitchen workers sowing foodborne epidemics, from *Shigella* to *Campylobacter* to hepatitis A viruses. Though shellfish from polluted waters are a major source of Norwalk virus, a nasty ailment that begins with sudden projectile vomiting, so are infected kitchen workers who are less than meticulous—such as the bakery worker in Minnesota in 1982 who used his bare hands and arms to stir 76 liters of butter-cream frosting, an unconventional culinary technique that made 3,000 pastry lovers sick. In 1988, the largest home-grown epidemic of *Shigella* came to light when 21 players on the Minnesota Vikings football team suddenly became ill after a charter flight to Miami, a trip on which they had partaken of cold meat sandwiches. After media coverage of the outbreak, hundreds of calls poured in to the Minnesota Department of Health from other air travelers complaining that they, too, had gotten sick. Investigators ultimately tracked the outbreak to 219 airline flights depositing travelers in 24 states, the District of Columbia, and four other countries—jaunts on which as many as 35,000 travelers ate contaminated sandwiches. One or more employees in the flight kitchen—who couldn't afford to lose a day's pay—had worked while sick with diarrhea from shigellosis.

It's these hidden, unsuspected sources of food poisoning—from long-contaminated meat and produce, from infected food handlers, from unpasteurized apple and orange juices—that have persuaded food safety advocates that fallible humans simply can't act as the bulwark against dangerous pathogens. In the United States, at least 12 federal agencies have a hand in food safety, fiefdoms that sometimes work in efficient synchrony but often in a kind of all-thumbs opposition. Though the system has become a sprawling mess, it did make sense during the era of malfeasance when it was created. In 1906, Congress passed the Pure Food and Drugs Act, which established the duty of the federal government to regulate foods other than meat and poultry, and to prohibit the interstate sale of food that was misbranded or adulterated with chemical preservatives—a structure that's now the Food and Drug Administration. And in response to Upton Sinclair's 1905 novel *The Jungle*, which graphically depicted Chicago's meatpacking industry, lawmakers passed the Meat Inspection Act of 1906. That law set sanitary standards for slaughter of animals and for meat sold in interstate commerce and led to daily inspection in slaughterhouses, using "organoleptic" means—sight, smell, touch—to ferret out problems.

Back when the nineteenth century became the twentieth, of course, "clean" didn't mean free of microscopic pathogens, since many were not known at the time. Because of this technological lag, the public and the courts later came to consider disease-causing organisms in meat and poultry an unavoidable risk. As late as 1974, in a case in which the American Public Health Association sued the U.S. secretary of agriculture for violations of the Wholesome Meat Act, a U.S. court of appeals endorsed this fatalism, writing that *Salmonella* and other organisms were "inherent" in meat and that it was the consumer's responsibility to work around those hazards: "American housewives and cooks normally are not ignorant or stupid and their methods of preparing and cooking of food do not ordinarily result in salmonellosis." Not until 1994 did a district court break a new precedent, declaring that "*E. coli* is a substance that renders 'injurious to health' what many Americans believe to be properly cooked ground beef" and therefore "fits the definition of an adulterant."

Shaped by case law and political wire-pulling, the USDA and FDA have dramatically evolved from their origins. The USDA is now responsible for monitoring meat, poultry, and commercially processed egg products, with a policing force known as the Food Safety and Inspection Service (FSIS). The FDA is charged with ensuring that all other foods are safe, nutritious, sanitary, wholesome, and honestly labeled. This jigsaw puzzle sometimes doesn't fit. For example, while the FDA oversees shell eggs, the USDA has authority over egg products. While the FDA oversees plants producing cheese pizza but rarely inspects them, the USDA has jurisdiction over plants producing pepperoni pizza and inspects them every day. And while the USDA is responsible for meat and poultry safety, its inspectors can look only for microbes that cause animal—not human—disease. The two agencies also share monitoring of imported foods, again with very different approaches. While the USDA must certify an exporting country's meat and poultry inspection program, the FDA has no such power and instead relies on a tiny staff to do physical inspection and chemical analysis of food at ports of entry—inspections that can't possibly keep pace with the huge rise in imports. The FDA inspects less than 1 percent of the foods and ingredients imported from other countries.

Today, the centerpiece of the federal food safety system is the HACCP program—for Hazard Analysis Critical Control Point. HACCP (pronounced *has´-sip*) is a system of process control that evolved out of methods designed to protect astronauts from food contamination. The model is an eminently rational one. It identifies critical contamination points all along the path from farm to fork—in shipping, processing, wholesaling, retailing, and kitchen handling—eliminates those hazards, monitors the effects, and then sets even more stringent standards. But it may hand over too much responsibility to industry to police itself. In 2000, the General Accounting Office, a research arm of Congress, issued a report recommending that the government tighten enforcement of sanitation standards in meat and poultry plants, and give the USDA the authority to levy fines against plants that don't meet these standards. The GAO also estimated that 85 percent of foodborne infections comes from fruits, vegetables, seafood, and cheeses: items regulated by the thinly staffed FDA, an agency with only a tenth of the inspection force of the USDA.

Critics of American food safety policies have insistently called for a single independent government agency to replace the current patchwork of bureaucracies and fiefdoms that are often at cross-purposes. CDC officials privately grumble about the USDA's and FDA's foot-dragging and the bureaucratic hurdles that spring up the instant an outbreak investigation is launched. "Until a few years ago, the attitude of the Department of Agriculture was that if a pathogen didn't make the animal sick, it wasn't their business—it was your business," says Carol Tucker Foreman, a former USDA official who now heads the Safe Food Coalition, an advocacy group in Washington, D.C. "You have diffuse authority and a political structure that says: Do not impose any cost on the farmer. You don't go tell the American farmer that he has to rebuild his stock pens, redo his water supply, alter completely the way he houses and ships animals to market. Look at the people who make up the House and Senate agriculture committees: Nobody's on those committees because his or her first interest is food safety. They're on those committees because their first interest is to make sure farmers make a profit."

Meanwhile, industry and regulatory officials counter that CDC investigators operate with a naive, narrow-minded zeal. Egg industry officials, for instance, complain that the Clinton administration and the CDC focused attention on eliminating *Salmonella enteritidis* in eggs only because SE outbreaks almost always originate with that single food, making it a conveniently neat target—unlike *E. coli* O157:H7 or other forms of *Salmonella*, which pop up all over. "Some of the people that we interact with from CDC . . . say things . . . without ever having personally been on a farm," observes the USDA's program leader for the National Animal Health Monitoring System, Nora Wineland. "I would take it with a very large grain of salt." So deep is the antagonism at times between the federal agencies involved in food safety that Wineland has even wondered aloud whether the many cases of disease that CDC officials are convinced are foodborne actually come from food at all. It's perfectly plausible, Wineland says, that someone could get a *Salmonella* infection by preparing food on a kitchen counter where the family cat may have just been prowling. "And where's that cat been?" Wineland asks. "When you try to think about where people could pick up these bugs, the possibilities in my mind seem nearly endless." It's an argument that makes CDC investigators see red.

The finger-pointing goes in all directions. Public health officials blame farmers and industry for sanitation problems that they believe perpetuate the chain of infection. Farmers blame slaughterhouses and processing plants for contaminating their products. Microbiologists studying animal diseases insist that fresh produce is the biggest threat today. And the tension between blaming consumers for foodborne disease and blaming corporations and government never goes away. "It's the blame game," says Mike Osterholm, adding that rhetoric doesn't go far. "Today, if I gave some of these advocacy groups a billion zillion dollars and told them to raise cattle and bring them to market and get them to the consumer without *E. coli* O157:H7, they couldn't do it. And if I gave a billion zillion dollars to a meat packing plant and told them, 'You can't have any *Listeria,* period—zip, zero,' they could do a lot to minimize it but they couldn't reduce it to zero." Osterholm wonders, in fact, if the problem could ever be dispelled at the source. "The bottom line is, if everybody did their job perfectly well, given the level of technology that's readily available, you'd still have problems." Food recalls, he adds, almost always come too late. "It's like shutting the barn door after the cow is out." Unlike many public health experts, Osterholm has pushed hard for a last-stage technical fix: irradiation of ground beef, poultry, and eventually of fresh produce, all of which he likens to the commonsensical pasteurization of milk.

Ultimately, preventing foodborne infections will take scientific and political courage—sparked, most likely, by personal stories. In 1993, Alexander Thomas Donley, age six, was one of four children who died from eating a Jack in the Box hamburger. *E. coli* O157:H7 first made him curl into a fetal position from abdominal cramps. Then, one after another, his organs failed. Screams of pain were followed by silence as toxins liquefied his brain. He suffered tremors and delusions and finally a massive seizure. His body swelled as his kidneys shut down. "I was so horrified and so shocked and so angered by what happened to him," says his mother, Nancy Donley, now the president of Safe Tables Our Priority, or S.T.O.P., a Chicago-based advocacy group. "I had no idea that there was any problem in our food supply. I loved my child more than anything in this world. And then to find out that he died because there were contaminated cattle feces in a hamburger. And to find out that had been recognized as a problem for a while. Why hadn't it been fixed?"

Donley believes in the value of teaching consumers to handle food safely, but she doesn't think education is enough. The best way to prevent foodborne illness, she says, is to mount a strong attack on the farm and then vigilantly follow up every step along the path to the table. It's an approach that goes against the grain of many political interests. Not long ago, testifying before a congressional subcommittee, a senator reminded her, as have countless industry and government officials over the years, that the United States has the safest food supply in the world. Donley stared at him and said, "Senator, I beg to differ with you."

Chapter 4

Superbugs

N ot long before antibiotics entered our lives, staph in the bloodstream killed 90 percent of its victims. A man who nicked him
self shaving could die from erysipelas, a strep infection. Children lost their playmates to scarlet fever, meningitis, osteomyelitis. Bacterial pneumonia, the leading cause of death, killed a third of its victims. Tuberculosis patients were advised to rest and seek clean air,
because there was nothing else medicine could offer. In parts of the
United States, 10 percent of women died in childbirth. Rocky Mountain spotted fever killed 20 percent of its victims. The state-of-the-art
treatment for syphilis was mercury ointment. Gonorrhea had no cure.
A blindfolded visitor led through a hospital could identify the surgical
ward by the stench of rotting flesh.

Today, antibiotics save tens of thousands of lives yearly in the

United States alone. Without these therapies, modern invasive medicine would come unhinged. Daily life would be an obstacle course of fear. Are we about to return to those times?

In May 2001, Mary Jane Ferraro was working at her desk when a lab supervisor stepped into her office. The woman had worked with Ferraro for 25 years, and they trusted each other's judgment completely. "You won't believe this," the supervisor began. Ferraro, who directs the clinical microbiology laboratories at Boston's Massachusetts General Hospital, listened hard. Her job is to make sure that one of the most storied medical institutions in the world—the place where patients go for help when the best outside doctors are stumped—pins down the most unexpected and difficult-to-discern agents of infection. An outgoing woman with a reddish-brown pageboy and a lyrical laugh, Ferraro soon headed down the hall to the lab, a complex of neutral-colored rooms with computers and microscopes and boxy incubation machines that warm bacterial cultures to human body temperature. The supervisor handed her a clear, round plastic plate. Growing inside were creamy yellow colonies of *Staphylococcus aureus.*

S. aureus is a virulent and aggressive pathogen, endemic in virtually every hospital in the world. On this May morning, the colonies of staph on the culture plate told a frightening story. In a circular arrangement on the plate were little round filter-paper disks, each saturated with a different antibiotic. If the antibiotic was effective against the staph, a wide clear ring formed around the disk, evidence that the nearby bacteria had been killed. If the bacterial colonies grew closer to the edge of the disk, it meant the drug was useless. Ferraro peered at the tiny disk containing linezolid. Barely a year earlier, to much fanfare, the Food and Drug Administration had approved the drug as the "last-resort" antibiotic for multidrug-resistant *S. aureus*, the most dangerous source of hospital infection. Here was the antidote doctors could supposedly turn to when every other medication failed. Ferraro couldn't believe her eyes. The creamy colonies of bacteria grew undeterred. It was the first evidence of linezolid-resistant staph in the world. She felt an odd mix of emotions: competitive pride that her laboratory had caught it, and terror that the infection had already spread to other patients.

Living Pharmaceuticals

Three millennia ago, the Chinese applied spoiled soybean curd to cure carbuncles, boils, and other skin infections. Mayan Indians roasted green corn and let it rot to produce a preparation that relieved skin ulcers and intestinal infections. Renaissance-era Europeans placed loaves of bread on rafters to grow moldy; this staff of life was sliced and mixed with water to make a pasty dressing for wounds. In 1640, the apothecary of London, moonlighting as the King's herbalist, prescribed a fungus to treat infected wounds. All were applying living pharmaceuticals, now known to be antibacterial toxins produced by fungi.

In the nineteenth century, bacteriologists shed new light on these time-tested practices. They discovered that, in the lab as in nature, the fittest microbes survived by killing other microbes. As Louis Pasteur observed, "In the inferior organisms, still more than in the great animal and vegetable species, life hinders life." The practical implications were obvious. "If the study of the mutual antagonisms of bacteria were sufficiently far advanced," two scientists wrote in 1885, "a disease caused by one bacterium could probably be treated by another bacterium." At the turn of the last century, German physician and bacteriologist Paul Ehrlich likened the immune system's antibodies to magic bullets that aimed straight for their bacterial targets while leaving everything else in their paths unscathed. Why not, Ehrlich wondered, find chemical bullets that were equally potent and benign? As a 1925 editorial in the *Lancet* prophetically observed, "Medicinal properties attributed by tradition to certain fungi may possibly represent an untapped source of therapeutic virtue."

In the warm September of 1928, Scottish bacteriologist Alexander Fleming returned to his London laboratory after spending two weeks at his country home. As Fleming perused a set of glass petri dishes that he had inoculated with bacteria before embarking on his holiday, his eye stopped. In one, a green contaminating mold had unexpectedly taken residence. Immediately surrounding the mold, the growth medium looked clear. Only around the far edges of the dish grew the colonies of *Staphylococcus aureus* that Fleming had seeded in the medium. A closer look revealed what Fleming called "ghosts" of the staph bacteria—transparent remnants left when the mold "lysed," or disintegrated, them. To Fleming, this was curious indeed. Just a few years before, he had identified the enzyme lysozyme, a component of tears and mucous fluids that kills non–disease-causing bacteria. But he knew that *Staphylococcus aureus,* one of the most dominant and fearsome groups of bacteria—to this day—was notoriously hard to lyse. "Obviously," he would later write, "something extraordinary was happening."

In a paper published the next spring, he identified the green contaminant as the fungus *Penicillium notatum,* from which penicillin got its name. Ultimately, it would be shown to destroy no fewer than 89 different pathogenic bacteria.

Fleming's new germicide was unstable and difficult to produce. But the chase was on. Prontosil, first prepared by the German dye company I.G. Farbenindustrie in 1932, ushered in the sulfonamides, or sulfa drugs. These "bacteriostatic" agents didn't actually kill microbes but rather checked their growth so that the immune system could finish them off. Ten years after Fleming's initial find, a team of Oxford scientists purified his mold juice. In 1940, having inoculated eight mice with deadly doses of streptococci, they injected four with penicillin as well. By the next morning, only the treated animals had survived.

It took a while to produce these magic bullets in bulk. In a legendary 1941 case, an English policeman had scratched the corner of his mouth while pruning roses. He developed staphylococcal septicemia, blood poisoning that left abscesses over all his organs. Administered over five days, penicillin pulled him back from the brink of death. But the purified yellow powder was dwindling, and when the drug ran out, the patient relapsed and died.

To save the lives of Allied soldiers, the U.S. government quietly underwrote the production of larger quantities. After D-Day, penicillin was released for civilian use. It became as common as candy, mixed in over-the-counter salves, throat lozenges, nasal ointments, even cosmetic creams. The drug and its synthetic derivatives transformed the medical landscape, dramatically cutting deaths from staph infections, sepsis after childbirth, pneumococcal pneumonia, ear infections, and bacterial meningitis. As Stuart Levy writes in *The Antibiotic Paradox,* "It was as if Prometheus had stolen fire from the gods."

But just as soon as antibiotics were discovered, so was antibiotic resistance. In a 1945 *New York Times* interview, Fleming warned about the evolutionary selection of resistant bacterial strains. He had seen it in his own lab. When he grew susceptible bacteria in ever-increasing amounts of penicillin, some flourished and took over. "There is probably no chemo-therapeutic drug to which in suitable circumstances the bacteria cannot react by in some way acquiring 'fastness' [resistance]," he warned. As early as 1946, a London hospital reported that 14 percent of *Staphylococcus aureus* strains taken from sick patients could stand up to penicillin. Three years later, that figure leaped to 59 percent. The term "hospital germ" was coined, reflecting the sense that certain microbes thrived in the dense presence of antibiotics—a far cry from the sanitarian view that dangerous infections lurked in the street or at work or on public transportation. By 1955, when most countries curbed the use of penicillin to prescription-only, it was already too late. Today, just under 100 percent of *Staphylococcus aureus* are penicillin-resistant. And that's just the beginning.

As scientists began prospecting for new antibiotics, their search led them to the very ground beneath their feet. As far back as the 1880s, it had been conjectured that the reason soil wasn't the source of perpetual epidemics—despite receiving human and animal wastes and corpses—was because of an invisible, ongoing warfare between microbes. Today, we know

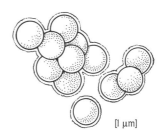

[I μm]

Staphylococcus aureus
Agent of skin infections,
bloodstream infections, pneumonia

that each teaspoon of soil may contain thousands of different species, a more biodiverse menagerie than the earth's entire collection of mammals. Whenever resistance emerged, researchers found a new class of drugs or chemically tweaked existing compounds. In 1944, Selman Waksman, a Ukrainian emigré to the United States and renowned soil microbiologist, systematically screened thousands of soil samples before happening upon a species named *Streptomyces griseus.* From this was extracted the first aminoglycoside: streptomycin, an early cure for tuberculosis. The genus of bacteria known as actinomycetes, which include the soil-loving *Streptomyces,* has since spun off well over half of the antibiotics used today, including the aminoglycosides, the macrolides, and the tetracyclines. The cephalosporins, refined versions of which are used today to prevent a wide range of hospital infections, were discovered in 1945; chloramphenicol, which cured typhoid fever and typhus, came in 1947; chlortetracycline, which cured pneumonia, in 1948; erythromycin, a broad-spectrum drug that stopped organisms such as *Neisseria* and *Haemophilus,* in 1950; vancomycin in 1955; methicillin in 1960; gentamicin in 1963; ciprofloxacin, a low-toxicity fluoroquinolone, in 1983.

Doctors wanted to be able to prescribe something even if they didn't know what organism was causing a patient's infection. So the pharmaceutical industry widened the killing swath of each new drug. As the antibiotic age unfolded, these categories would become important. Narrow-spectrum drugs such as penicillin G affect mostly Gram-positive bacteria, such as staphylococci, streptococci, and enterococci, so named because their single cell wall takes up a special stain used to differentiate microorganisms, and turns dark purple-blue. Gram-negatives, a group that includes gonococci, meningococci, and many intestinal bacteria, have a three-layer cell wall and so do not absorb the dye, showing up on a lab slide as a bright pink-red. Drugs that killed Gram-positive organisms often didn't kill Gram-negatives. Not until researchers perfected "broad-spectrum" antibiotics and later even more widely effective "extended-spectrum" agents, were they able to treat both with the same drug. While on its face this development typified the golden age of therapeutics, it would have unforeseen downsides. The further antibiotics extended their reach, the wider the bacterial insurgency.

It was 1967 when investigators isolated the first penicillin-resistant pneumococcus, the highly invasive organism that causes middle ear infections, pneumonia, and meningitis. Found in a healthy three-year-old boy in a remote village in Papua New Guinea—and probably bred through a misguided effort to prevent such infections by injecting residents with low-level penicillin—it was seen as a fluke. Looking back, that judgment could be interpreted as myopia, wishful thinking, willful ignorance, or just plain laziness. It's also true to form. The history of the antibiotic era is littered with blasé dismissals of newly resistant species as exceptions. Ten years later, as if on cue, drug-defying pneumococcus surged, newly equipped with biochemical weapons against several classes of drugs. Resistant strains spread to South Africa in the 1970s, to Spain and Israel in the 1980s, and to this country in the early 1990s.

In the 1970s, other common infections also began a portentous shift. Within a year after the introduction of methicillin, a semisynthetic version of penicillin, methicillin-resistant strains of staph showed up. These became epidemic in Europe before vaulting to the United States. *Haemophilus influenzae*, a major bacterial cause of ear infections and meningitis in children, became penicillin-resistant—a turn of events revealed when two infants who had attended the same Maryland daycare center died from ampicillin-resistant meningitis. Gonorrhea caused by penicillin-resistant strains of *Neisseria gonorrhoeae* radiated out from brothels in Vietnam where women received prophylactic doses of the drug, no doubt intended to protect the troops more than the women. For public health officials, all these trends were wake-up calls, signaling that common infections contracted outside of hospitals could soon become untreatable. In the early 1990s, a multidrug-resistant strain of tuberculosis killed nearly 500 people in New York City, mostly AIDS patients. Just as the outbreak seemed to diminish, TB reared up elsewhere in the world, from China to Siberia to Puerto Rico, now resistant to four of the five drugs used to treat it.

With global travel, public health experts worry that these divergent strains will merge and become incurable.

With each passing year, what a researcher in 1957 called the "bugbear of drug resistance" shadowed every antibiotic success. By the mid-1990s, the flow of replacement antibiotics slowed to less than a trickle. As infectious diseases resurged in this country, the media warned of "Andromeda strains" and "superbugs." Those headline prophecies would soon come true.

Resistance Forces

The more we try to eliminate deadly bacteria with new drugs, the better bacteria get at dodging medicine's magic bullets. So goes the "antibiotic paradox," to use the succinct phrase of Stuart Levy, director for the Center for Adaptive Genetics and Drug Resistance at Tufts University School of Medicine, and for more than 20 years a cautionary voice. Every antibiotic ever used has hastened its own futility by triggering Darwinian selection. When an antibiotic attacks a group of bacteria, cells susceptible to the drug will perish. Those that can resist the drug, whether because of genetic mutations or because they have acquired protective genes from other bacteria, survive. Facing less competition for space and nutrients—antibiotics having killed off their natural competitors—these resistant cells multiply. The more they spread, the more they add to the pool of resistance genes in all bacteria, raising the odds that these genes will jump to more and more disease-causing bacteria.

In the laboratory, resistance plays out neatly and in full view. Outside the laboratory, the process is wild and sloppy and mysterious. Resistance genes flow in all directions. And though their travels cannot be tracked in real time, they can be extrapolated. Give your teenager antibiotics for acne, and you might soon develop drug-resistant staph on your own skin. Treat a toddler with cephalosporins for an earache, and soon a majority of the daycare center's young clientele (and their parents) will be suffering resistant pneumococcal infections. Feed a flock of chickens low doses of tetracycline, and within days not only will their coopmates harbor intestinal bacteria highly resistant to the drug, but so will their human handlers and their human diners. Pull back the camera even further, and it becomes clear that as all the earth's bacteria became bathed in antibiotics—millions of tons over the past half century—they have, to varying degrees, formed resistance. Resistant bacteria have even invaded U.S. waterways, from the Rio Grande to rural streams; at some sites, half of the waterborne bacteria resist ampicillin, tetracycline, and vancomycin.

Antibiotic resistance develops when two ingredients come together: resistance genes and antibiotic use. If bacteria in a community—a community being a home, a daycare center, a school, a hospital, or a city—don't have genes that enable them to withstand an antibiotic, that drug will simply mow them down, end of story. If the bacteria do have resistance genes but are not exposed to the antibiotic, they will have no selective advantage against their competitors and will eventually dwindle in number. Or at least, that's the theory. In reality, things get a little trickier. Bacteria exposed to one drug can sometimes develop resistance to a whole raft of unrelated agents. And long after they've stopped being exposed to an antibiotic, bacteria will sometimes hang on to their genetic defenses.

But the general theory holds true: antibiotic resistance requires resistance genes plus exposure to an antibiotic. The more widely an antibiotic is used, the more resistance shows up in many different bugs.

The consequences can be seen all around us. A child with an ear infection that in the early 1990s would have been instantly cured by penicillin may now need two or three or four courses of different drugs. A new mother may no longer die of "childbed fever," but she might well contract a drug-resistant urinary tract infection that keeps her in the hospital for another day or so. In the 1990s, a Boston hospital twice had to construct a new neonatal intensive care unit (ICU) because its newborns had been colonized with highly drug-resistant staph. In 1999, a New Jersey schoolteacher who went in for surgery to remove small growths from her sinuses wound up with a staph infection that kept her out of the classroom for a year. Every day, hospital patients find themselves alone in a room with a bright yellow "precautions" warning on the door, declaring that visitors must don latex gloves and other protective accoutrements. What's not explicitly stated: the patient has an incurable infection that could spread to other patients.

If the trends keep up, our most commonplace infections could turn into life-threatening illnesses. "It's probably the biggest public health threat to the United States," says J. Glenn Morris Jr., chairman of the department of epidemiology and preventive medicine at the University of Maryland School of Medicine. "Antibiotic resistance is something that directly affects everybody in this country, in a way that no single disease does."

"For many common infections, we're losing the drug of choice," adds David Bell, who monitors antimicrobial resistance for the Centers for Disease Control and Prevention. Usually, there's a second or third drug choice—but those medicines may be less effective or more toxic or may have to be given by injection instead of by mouth. And then there are organisms for which the rising tide of resistance has swept away medicine's second or third choices. In those cases, says Bell, "we're skating just along the edge."

Sick Beds

Most resistant infections strike people in hospitals, nursing homes, and long-term-care facilities. That's no surprise. These are places where antibiotics are used most intensively and so are the natural proving grounds for resistant infections. Because hospitals have been downsized at the same time as intensive care beds have increased, many facilities have in effect become giant intensive care units, where the sickest, most vulnerable patients are concentrated. Five percent of all U.S. hospital in-patients—about two million people annually—contract infections in hospitals. About 70 percent of those infections are caused by microbes that resist one or more antibiotics. Of those, 30 to 40 percent of the infectious organisms rebuff the drugs doctors would choose first. For the seriously ill, the immune-compromised, the very young, and the very old, any delay in curing a resistant infection raises the risk of death. Each year, according to the World Health Organization, drug resistance contributes to more than 14,000 U.S. hospital deaths.

If resistant pathogens could dream of paradise, their reveries would look like the inside of a hospital. Take *Staphylococcus aureus.* Staph homesteads in the human body, living benignly on the skin of 20 to 25 percent of us, often in our nostrils, armpits, or groin. The rise of antibiotic-resistant staph has gone hand in hand with the invasionary forces of medical progress, especially central venous catheters, those long plastic tubes that snake all the way to the heart. In intensive care units, oncology and surgical wards, and dialysis patients, these tubes are commonplace because they make it easier for nurses to give medications and blood infusions. The linezolid-resistant *S. aureus* that Mass General's Mary Jane Ferraro confronted was a typical example; it had come from an 85-year-old dialysis patient who had a fixed catheter in his abdomen. Starting in the 1980s, having a "plastic" in the body became a major risk factor for resistant infections from either *S. aureus* or another common type known as coagulase-negative staph. Bacteria—either from the patient's own skin or from the hands of a health care worker—cling to catheters and other prosthetic material. Patients with open wounds such as bedsores also develop staph infections. Once staph gets into the bloodstream, it can cause heart valve, blood, and bone infections, sometimes leading to septic shock and death.

In the United States, July 1997 was a turning point for public health officials who monitor hospital antibiotic resistance. That month, a 59-year-old Michigan man became infected with a strain of *Staphylococcus aureus.* Long ago, of course, *S. aureus* had stood up to penicillin. Now it defied other drugs as well. It defied methicillin, the 1960s synthetic replacement for penicillin. That act of insurrection earned the bacterium a name—MRSA, for methicillin-resistant *Staphylococcus aureus*. But its recalcitrance didn't end there. MRSA had gone on to resist chloramphenicol, ciprofloxacin, clindamycin, erythromycin, gentamicin, imipenem, tetracycline, trimethoprim, and others. By the time it reached the Michigan man, MRSA—which now in effect stood for multidrug-resistant *Staphylococcus aureus*—reliably caved to one medication only: vancomycin, the drug of last resort. In the laboratory, the smallest concentration of vancomycin that killed the bacterium was 0.5 micrograms per milliliter of solution. This was known as the MIC, the Minimum Inhibitory Concentration.

This time, it didn't work.

Even at 16 times that dose, the drug didn't work. In the protocol of microbiology labs, that bumped up this MRSA to a new level of danger, and a new acronym: vancomycin-intermediate *Staphylococcus aureus*, or VISA. (Today, the less euphonious handle is GISA, with the G standing for Glycopeptide, the class of antibiotics that includes vancomycin.) In practical terms, intermediate resistance is a gray zone. Maybe high, toxic doses of vancomycin would have cured the infection—or maybe not. So far, clinical evidence suggests that VISA infections are impervious to vancomycin therapy. Doctors removed the stomach tube that had caused the man's infection in the first place—he was a diabetic with kidney failure, and used home dialysis to clean his blood—and the infection eventually cleared, though the patient soon died of his underlying disease.

Only a month later, another GISA patient appeared: a 66-year-old New Jersey man, a diabetic with a staph infection in his bloodstream. He, too, survived the GISA but died of other causes. By 2000, three more elderly and chronically ill patients with GISA succumbed to the infection.

This was what public health officials had been dreading, ever since the world's first case in a Japanese baby in 1996. The infections were isolated, developing anew in each patient, through mechanisms of natural selection all too familiar to doctors. Luckily, they didn't spread to other patients. But public health officials feared that someday, as MRSA had in the 1980s, these new infections would race through hospitals. GISA was one step away from becoming untreatable staph. And untreatable staph would be, if not a full-blown reenactment of the preantibiotic age, at least a convincing one-act play. "Once you have the vanco resistance, and then linezolid resistance, there are no antimicrobial agents that could be used right now," says Mary Jane Ferraro. "Could the problem spiral out of control? Yes. It is a realistic fear."

In some hospitals, upwards of 70 percent of *S. aureus* infections are of the multidrug-resistant MRSA strain. To cope with these infections during the 1980s, and with an increasing number of infections with coagulase-negative staph and with a stubborn bacterial pathogen appropriately known as *Clostridium difficile*, hospitals cranked up their use of vancomycin, sometimes 200-fold. The irony is, doctors had always considered vancomycin a so-so drug. "I hate to see vancomycin termed 'the drug of last resort,'" says Glenn Morris. "It's a second- or third-string antibiotic, but if everybody else has already fouled out, that's all you've got left. Vanco is not a superstar." Nevertheless, this bench-warmer selected not only for resistant staph, but another frightening bacterium as well.

That other fearsome microbe is VRE, for vancomycin-resistant enterococci. First appearing in the United States in 1989, VRE infections rose 20-fold over the next four years. The bacterium owes its success to an ecological stroke of luck. The vancomycin prescribed for resistant staph rearranged the landscape inside patients' intestines. Enterococci—a part of the normal microflora of the human intestine and until recently considered to be microbial milquetoasts—began to resist vancomycin. And when they escape the intestine, they become unpredictable. Entering the bloodstream through an IV line, a urinary catheter, or any breaks in the body's defenses, VRE attacks wounds, heart valves, the urinary tract, and can lead to life-threatening infections. In a Pittsburgh hospital outbreak, an epidemic of VRE killed more than 30 patients in a liver-transplant unit. At risk for VRE is any hospital patient who has received an organ transplant, is on hemodialysis, takes an immune-suppressing drug, or has simply been weakened by fighting off infection. The *last* last-resort drug to be approved for VRE infections was linezolid, which goes under the trade name Zyvox. Barely a year after its approval, John Quinn, a physician at the University of Illinois at Chicago, published the first report of linezolid-resistant enterococcus; it emerged in five chronically ill patients, four of whom had undergone transplants.

VRE's infamy stems not only from the damage it does today but also from a deadly supporting role it's poised to play in the future.

MRSA and VRE are bad enough. MRSA *plus* VRE is a calamity. Researchers fear that because VRE thrives in the same critically ill patients—often in the same wounds—as does MRSA, staph could steal enterococci's vancomycin-resistance genes. The result: the most widespread hospital infection would be untreatable. (In England, researchers successfully performed this DNA transfer in the lab—and claim to have destroyed the resulting organism.) At the CDC, William Jarvis, who directs investigations for the Hospital Infections Program, worries that most doctors dismiss VRE as "wimp bugs," less invasive and deadly than *S. aureus*. "They fail to recognize that VRE is colonizing the same populations of high-risk patients that MRSA colonized in the past. We're getting this larger and larger pool of patients who are repeatedly admitted to our hospitals, repeatedly placed on antibiotics," he says. "It's a time bomb. Sooner or later, that genetic transfer will occur."

If *S. aureus* resisted every last-resort antibiotic, a shadow would fall across medicine. Surgeons would think twice before doing elective procedures, from hip replacements to cyst removals, for fear of life-threatening complications. More people would die in hospitals: intensive care patients, surgical wound patients, dialysis patients, transplant patients, even otherwise healthy patients who happen to get a staph infection from an IV. "Clusters or outbreaks of infections could lead to closings of wards, floors, even hospitals," says Mitchell Cohen, who directs the CDC's division of bacterial and mycotic diseases. At least a quarter of the 50,000 serious hospital staph infections each year in this country would be fatal. Health care workers could become potential carriers and a threat to patients.

Staphylococci and enterococci are Gram-positive. In the 1990s, after a brief hiatus, Gram-negative organisms ominously rose in hospitals. They have impressive names: *Pseudomonas aeruginosa, Acinetobacter baumannii, Serratia marcescens, Burkholderia cepacia, Stenotrophomonas maltophilia,* and so on. But like MRSA and VRE, Gram-negatives are the dark side of medical progress. Since many of these newly ascendant Gram-negatives are water-loving bugs, they have made themselves at home in respiratory and gastric tubes, sink drains, even flowerpots. "The reason nobody talked about *Stenotrophomonas* a few years ago was that there weren't many people who lived on respirators as long as people do now," says Robert Arbeit, an infectious disease specialist at Cubist Pharmaceuticals. "You've got this tube and this open airway and this mucosal surface that's exposed to the environment. Patients get infected and then they get courses of antibiotics, which promotes resistance." Some Gram-negative organisms resist every available drug. As a result, patients get second surgeries or other procedures. Or they die.

With hospitals whipping up armies of totally and near totally drug-resistant bacteria, what worries public health officials is that these armies may venture outside. The boundary line between hospital and hearth has been blurred lately. Sick patients often go home with intravenous lines dangling from their arms even before an infection is cured. They may trek three times a week to a clinic for dialysis. Chronically ill patients circulate between hospitals and ambulatory care centers and nursing homes. Medical progress, in other words, actually puts more people at risk of infection. "We do surgeries. We do transplants. We knock out bone marrow. We do aggressive chemotherapy," says Glenn Morris. "All of those procedures are dependent on our ability to kill bacteria, to eliminate the almost inevitable infection that follows when you knock out the body's immune system." And because resistant infections take longer to treat, more semi-recovered people spread those infections into their communities.

Most surgeries now take place in outpatient centers. Antibiotic-resistant bacteria are ferried to homes, schools, work, grocery stores—places they have never visited before. Health experts fear that hospi-

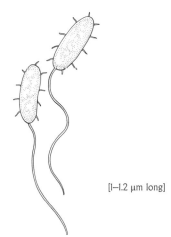

[1–1.2 μm long]

Pseudomonas aeruginosa
Agent of hospital-acquired pneumonia,
bloodstream infections

tals' stubborn drug-resistance genes will be taken up by otherwise benign community microbes. Today, if you're healthy and you're taking care of a sick relative just out of the hospital, chances are you won't get infected. But you could be the unlucky one, as more and more such bacteria—staph, enterococci, strep—strike out into the world.

***

How did hospitals become boot camps for resistant bacteria? Massive antibiotic use is one reason, of course. Just as important is what kind of antibiotics doctors prescribe in hospitals. Broad-spectrum antibiotics, which target both Gram-positive and Gram-negative bacteria, invite a wider counterforce of resistance. Doctors use broad-spectrum drugs to treat patients at high risk of quickly dying from infection—patients who can't wait 24 or 48 hours for a lab to culture an organism so the doctor can make a specific diagnosis. Ken Sands, a health care quality specialist at Beth Israel Deaconess Medical Center in Boston, describes a common scenario. "Someone comes into the emergency room deathly ill: sepsis, no blood pressure, about to die. It could be a college student with meningitis, an elderly patient with a urinary tract infection leading to sepsis, a cancer patient with an indwelling line for medications. The doctors will pile on six antibiotics, might as well throw the kitchen sink at it until you know what's going on."

There's a second reason resistant bacteria flourish in hospitals: hospital hygiene has grown lax. More than 150 years ago, the Austro-Hungarian physician Ignaz Semmelweis discovered that puerperal fever, a systemic infection that routinely killed women who had just given birth, spread on the hands of physicians who had also performed autopsies in the same clinic. Washing hands with chlorinated water solved the problem. Later, when the destructive power of microbes was well documented but miracle drugs hadn't yet arrived on the scene, hospital staffs took extraordinary care to prevent infections. When the germ theory gained credence, nineteenth-century reformers such as Florence Nightingale worried that it would undermine the achievements of sanitary science.

Apparently, she was right. Today, many medical personnel expect antibiotics to cover for sloppy hygiene. Yet, of all the paraphernalia in hospitals, workers' hands remain the most dangerous. VRE can survive on unwashed fingertips for 30 to 60 minutes. Whenever a nurse changes bandages on a weeping wound, then neglects to wash her hands or change gloves, MRSA gets a free ride. In 1997 and 1998, *Pseudomonas aeruginosa,* though not resistant, contributed to the deaths of 16 babies in an Oklahoma City hospital—about half the deaths due to contamination from nurses' long fingernails. With hospital staffs cut, nurses covering a large number of patients don't have the luxury of scrubbing their hands for 30 seconds between patients—the minimum it takes to wash VRE off their fingers with a good soap. A 1999 study from Duke University found that only 17 percent of ICU doctors properly washed their hands. Even if every hospital worker observed the utmost cleanliness so that the environment was not contaminated, it would eliminate only about a third of nosocomial (hospital-borne) infections, experts say; the rest will continue to arise when patients' own bacteria encounter germ-fighting drugs.

The Home Front

Not all antibiotic resistance is born in hospitals. The same month that the Michigan man came down with vancomycin-resistant staph, a healthy seven-year-old girl in Minnesota died from a different *Staphylococcus aureus* infection. Somehow, the bacterium had breached her skin, planted an infection in her right hip, and traveled through her bloodstream to her lungs, overwhelming her body's defenses. Staph is not supposed to kill healthy children. Her doctors were blindsided. Here was an infection that was highly unusual in the caseload of general practitioners: an MRSA strain that originated not in hospitals but in the community. Unlike most community staph infections, this one didn't respond to the typical first-line treatments, a group of drugs known as beta-lactams, which include penicillin and cephalosporins. When the girl's doctor prescribed these drugs, the staph continued to march through her body, unfazed. By the time her physicians turned to vancomycin, it was too late.

Six months later, in January 1998, a 16-month-old girl from North Dakota died of a bodywide staph infection. A year after that, a 13-year-old girl from Minnesota died of pneumonia caused by staph. The next month, in February 1999, a one-year-old boy from North Dakota died from staph pneumonia. And in January 2001, a strapping 21-year-old Minnesota college student died of a fast-moving pneumonia—just eight hours after his symptoms began. Like the first Minnesota girl, their infections were caused by an MRSA. Having survived the first round of drugs, the organism had time to find a target inside the body. In a common scenario, substances in the outer walls of the staph bacterium send the immune system into overdrive, blood pressure plummets, high fevers spike, and organs shut down. Staph toxins cause internal abscesses to metastasize. Once this cascade of events begins, vancomycin is powerless. In 1999, the CDC had reported the four young deaths as a warning to doctors that this new strain of staph had to be kept on a very tight leash.

The cases should not have come as a complete surprise. There had been scattered reports earlier of MRSA in injection drug users, on Native American reservations, and in daycare centers. After the children's deaths, CDC and state health department investigators uncovered hundreds of cases throughout the United States, mostly in children and teens—people whose infections, for unknown reasons, didn't progress with the same deadly ferocity as in the first young victims. Though these patients didn't die, many barely skirted danger; getting the wrong antibiotic was like getting no antibiotic at all.

If the staph that colonizes our skin starts to fend off many drugs, then the slightest skin infection—and skin infections are commonplace, the bread-and-butter of general practitioners—may someday require heavy-duty therapy. If you cut yourself chopping carrots or get an infected pimple, you could wind up with an IV in your arm. If ordinary skin infections resist more and more antibiotics, physicians will be forced to turn to the drugs of last resort. That could breed resistance in all sorts of bacteria in our bodies. And what would be left to treat the most life-threatening illnesses?

Another alarming home-grown organism is *Streptococcus pneumoniae*, better known as pneumococcus. To get a sense of its tenacity, consider the case of Hollie Mullin, who was only three weeks old when she suffered her first ear infection. Her mother, Gail Mullin, a schoolteacher and first-time mom in Olathe, Kansas, thought her newborn daughter needed antibiotics. Hollie's doctor wasn't so sure. "Her eardrums are pink—they're not red. Let's watch and wait," he told Gail. "If she starts acting fussy, give us a call." A day or so later, as Hollie wailed uncontrollably in the background, Gail phoned back the doctor and asked for amoxicillin. She wasn't sure whether colic or ear pain was causing Hollie's crying, but she couldn't stand to see her child in distress. "That started the ball rolling," she says now, in a voice tinged with embarrassment. She wishes she had never picked up the phone.

It was October 1997. In November, Hollie developed a second ear infection, and Gail pushed for another prescription of amoxicillin.

In December, Hollie suffered a third ear infection, and this time her doctor tried a stronger antibiotic: Augmentin.

On and on it went through 1998. Every month, Hollie came down with a new ear infection and was given either repeat doses of the same drug or was switched to an increasingly potent medication. Zithromax. Rocephin. Omnicef. Lorabid. Cleocin. By her first birthday, Hollie had gone through 17 separate courses of antibiotics. The germs in her ears kept coming back, emboldened with each confrontation.

The drug regimens continued into her second year. Sometime during this period, her doctors say, Hollie's ears became colonized with antibiotic-resistant bacteria. In June 1999, when she was 21 months old, Hollie came down with the worst ear infection yet—one that defied all the earlier drugs. This time, her physicians took drastic measures. While Hollie was under anesthesia, they threaded a catheter through a vein in her arm into her chest. Then three times a day for 14 days, while Hollie recuperated at home, Gail Mullin painstakingly administered to her daughter the antibiotic of last resort: vancomycin.

When the course of therapy was over, Hollie's doctors emphatically told Gail that there was nothing else left to try. The vancomycin had killed off all the remaining drug-resistant bacteria; now it was time for Hollie's immune system to regain strength. For the next six months, they warned her, Hollie absolutely had to stay well. Gail and her husband, Rob, kept their daughter home and shooed away sick visitors. Miraculously, Hollie's immune defenses rallied so effectively that her ear infections finally ceased. "We created our own monster," says Gail Mullin, "by using so many antibiotics that nothing worked."

Pneumococcus is the bane of parenthood, the microbe that causes five million middle ear infections in kids a year, as well as other childhood bacterial infections. In the days before antibiotics, most ear infections went away on their own if left untreated. Today in Europe, doctors ease them with pain killers rather than antibiotics; children return to the office if they're not better in two days. Ten years ago in the United States, treating pneumococcus with antibiotics was a snap. Daycare centers changed all that. Microbial cesspools where children on antibiotics are constantly exposed to each others' secretions, they became hot spots for penicillin-resistant pneumococcus. At first, the outbreaks mostly struck kids from the suburbs, who had the best access to medical care and were most likely to get prescriptions. The expensive antibiotics that these children received—the cephalosporins—were not terribly effective against pneumococcus, and quickly spawned resistance. By now, it's not only the comfortable who are afflicted; resistant pneumococcus has permeated the country.

The more antibiotics your child has taken in the past three months, the better his or her chances of carrying or getting infected by resistant pneumococcus. Depending on where you live, 25 to 40 percent of the bacteria resist penicillin and a growing list of other drugs. Pneumococcus is also starting to show "tolerance," as microbiologists put it, to vancomycin—that is, it can survive but not reproduce when exposed to the drug. Tolerance is just a step behind full-blown resistance. And if pneumococcus resisted vancomycin, it would be bad news. More children would suffer infections of the bone surrounding the ear. Some would go on to develop meningitis—a grim challenge to doctors because, since few drugs can reach the spinal fluid, delaying treatment for just a few hours can be fatal. Children wouldn't be the only sufferers. Pneumococcus also causes about two million cases of invasive pneumonia, mostly in older people, as well as more than 50,000 bloodstream infections a year. Those victims, too, would be more apt to die or suffer medical complications. And the effects of overtreating ear infections with broad-spectrum antibiotics may ripple even further. Drugs such as Augmentin may also be spawning resistant staph in daily life, such as the infectious agents that have killed children and young adults in the Midwest.

In 1998, the U.S. Senate held a hearing about antibiotic resistance. There, the most dramatic testimony came from a Mississippi mother named Angela Littlejohn. Her healthy 14-month-old daughter, Shauna, suffered a series of ear infections and was given a stepped-up arsenal of antibiotics. In 1989, Shauna slipped into a coma caused by drug-resistant pneumococcal meningitis. Today, Shauna remains blind and in a vegetative state, fed through a tube. "This nightmare still haunts me every day of my life," her mother told lawmakers. And it haunts public health officials as well: Shauna's was the first case of multidrug-resistant pneumococcus in the United States.

Cultivating Resistance

Not all resistant microbes spring from misguided human medicine. Farms are some of the most insidious sources of antibiotic resistance. As mentioned in Chapter 3, antibiotics are routinely fed in tiny amounts to farm animals—not to fend off disease, but to boost growth. And low-level use of antibiotics is a perfect way to foster resistant organisms. In recent years, livestock industry experts had estimated that 40 percent of antibiotics produced in the United States went to farm animals. In 2001, however, a report from the Union of Concerned Scientists, a Cambridge, Massachusetts–based environmental advocacy group, raised this estimate to a whopping 70 percent. Public health experts have long worried that farmers are squandering human lifesaving drugs on animals that are not even sick.

Take *Campylobacter*, the most common bacterial cause of food-borne illness. In 1995, American poultry farmers began using a fairly new class of drugs known as fluoroquinolones to treat respiratory infections in poultry. In people, these broad-spectrum, low-toxicity drugs are some of the most prized antibiotics today, because they are slow to breed resistance and are effective against some of the hardest-to-treat infections. Almost immediately after poultry farmers began dosing their birds with the medication, thousands of people who ate undercooked chicken contaminated with fluoroquinolone-resistant strains of campy themselves became infected with the drug-resistant bacteria. Before the drug was used, no *Campylobacter* specimens cultured from hospital patients had been resistant; today, nearly a fifth are, and the figure is sure to rise.

Even more dangerous than drug-resistant *Campylobacter* is resistant *Salmonella,* which is also present in poultry and meat. In human medicine, fluoroquinolones are the preferred treatment for invasive, and often life-threatening, *Salmonella* infections. Yet today, doctors are resorting to higher and higher doses of fluoroquinolones to treat *Salmonella*—a possible prelude to full-blown resistance. And now there's a frightening new wrinkle in treating the organism. In 1998, a 12-year-old Nebraska boy picked up a *Salmonella* infection from his family's cattle that was resistant to ceftriaxone—one of the cephalosporin class of antibiotics—as well as a dozen other antibiotics. Fortunately, he survived when doctors treated him with a combination of other drugs. But when this unprecedented case was reported in 2000, it terrified public health officials. Ceftriaxone is one of the few antibiotics that reliably kills most bacteria. And it is the drug of choice for children whose *Salmonella* infections have entered the bloodstream—a condition that kills about 1,000 Americans every year. Ceftriaxone is also the drug that doctors turn to when treating young victims; because of worries about bone growth, quinolones are not approved for children. Since 2000, more cases of ceftriaxone-resistant *Salmonella* in people have turned up. "This *Salmonella* is so multiresistant," says the CDC's David Bell, "there are no good drugs left that are approved for children." To history-minded physicians, the situation evokes futile attempts at the turn of the last century to treat typhoid fever, another *Salmonella* infection. Extrapolating from subsequent studies of patients, health officials calculate that tens of thousands of *Salmonella* cases each year are ceftriaxone-resistant. The clinical problem also touches on a moral quandary: ceftriaxone is not used as a growth promoter, but rather to treat sick animals. "It portends a dilemma," says the CDC's Fred Angulo. "Societally, what do you want to do: treat sick people or sick animals?"

Another foodborne infection is VRE—yes, the same bug that wreaks so much havoc in critically ill hospital patients. In this country, VRE isn't primarily foodborne; the organism is most often bred by massive vancomycin use in hospitals. It's a different story in Europe. Soon after farmers there began feeding avoparcin, a growth promoter related to vancomycin, to livestock in 1974, the animals developed vancomycin-resistant enterococci. (Because it may be a carcinogen, avoparcin never received approval in the United States.) In 1986, France found its first human patient with VRE. Within a few years, the bacterium spread throughout human intestinal tracts on the Continent. U.S. public health experts believe that at least some of the VRE organisms in this country may have come from Europe and then proliferated under the selective influence of vancomycin in hospitals here.

But while foodborne vancomycin-resistant enterococcus infections are uncommon in the United States, a similar chain of events is starting to happen here with another drug. For more than a quarter century, American poultry farmers have used the growth promoter virginiamycin in chicken feed. In chickens, the drug helped breed enterococci that are resistant to virginiamycin's human-use cousin, Synercid. Synercid is the other "last-resort" antibiotic, approved for humans in 1999. Yet as a frightening presentiment to the drug's potential downfall, more than half of grocery-store chickens carry bacteria impervious to this end-of-the-line human drug. People are picking up these resistant bacteria in their meals. At any one time, at least 1 percent of the U.S. population is carrying Synercid-resistant enterococci. Usually, these intestinal bacteria are expelled as food moves through the intestines, never causing a problem. But in the rare instance that such an individual enters the hospital—say, for a hip replacement—and happens to be treated with Synercid, the resistant enterococcal bacteria in the gut will go wild, threatening an infection that no antibiotic can quell.

One of the most frightening and enigmatic foodborne pathogens is a drug-resistant strain of *Salmonella typhimurium*. Known as Definitive Type 104, or DT 104, it defies five important classes of drugs in the United States; in Europe, where it surfaced in 1984, it thwarts seven. This monster resistance has helped it spread in cattle, because in animals that receive any one of these drugs, DT 104 gains an advantage. In the U.S., hundreds of thousands of people suffer DT 104 infections annually. Raw milk is a common culprit, the bacterium having infiltrated dairy herds. In 1997, for example, more than 100 Californians became sick from DT 104 in two overlapping outbreaks in Hispanic communities, where residents ate homemade Mexican-style cheese made from unpasteurized milk and sold by street vendors and specialty markets.

When scientists tried to figure out where this renegade came from, they were shocked. DT 104's resistance genes were a strange combination—so strange, they had never before been seen in *Salmonella*. Where they *had* turned up was worlds away: in Asian aquaculture, where fish have been regularly treated with antibiotics since the early 1980s. So how did they land in the American heartland? One theory holds that some of those Asian fish may have been ground up into fish meal, an international commodity often fed to pigs and poultry. Or DT 104's resistance genes may have found their way into animal breeding stock, perhaps through the rendered protein of other animals. However it happened, DT 104 appeared more or less simultaneously around the world in the 1980s, suggesting that the animals acquired these alien bacteria en masse.

In 1969, Britain's Swann Committee concluded that antibiotics used in human therapy or those that promote cross-resistance in people should be banned from animal growth promotion. Unfortunately, livestock producers hew to the position that whatever drugs they feed their animals are proprietary secrets. Besides, say industry officials, they need antibiotics to produce safe and affordable food. A 1999 report published by the Institute of Medicine and the National Research Council questioned this claim. Using the livestock industry's own estimates, the report calculated that if farmers quit using antibiotic growth promoters, the added costs would be less than $10 per American consumer per year. And a 2001 United States Department of Agriculture report showed that hog farmers actually lose money by giving pigs antibiotics that promote growth; while animals do fatten up more, the extra poundage expands overall supply and drives down market prices.

Poultry and livestock aren't the only creatures being dosed with drugs. Salmon, catfish, and trout on domestic fish farms get antibacterial drugs in the water. Honeybees get antibiotics in their hives. And each year, an estimated 300,000 pounds of antibiotic pesticides drift down on fruit trees and other crops to control or prevent bacterial infections such as fire blight. That disease is caused by the pathogen *Erwinia*, a bacterial cousin of *E. coli, Salmonella*, and *Shigella. Erwinia* now resists both streptomycin, an old drug, and tetracycline. Researchers don't know if the fresh fruit invitingly stacked in your supermarket is delivering drug-resistant genes to your intestines. According to microbiologist Abigail Salyers, both the use of untreated or partially treated water for irrigation or for washing vegetables, or the use of manure as a fertilizer for vegetables and fruits could contaminate food plants with antibiotic-resistant bacteria. Proving that no good deed goes unpunished, a 1993 study found higher levels of multidrug-resistant bacteria in intestines of vegetarians than in meat eaters. Whether carnivore or vegetarian, you cannot avoid the aftermath of antibiotics applied lower in the food chain.

Immaculate Misconceptions

Antibiotics have tilted the playing field in favor of resistant bacteria. By snuffing out the motley crew of benign, drug-susceptible bugs in and on our bodies, they cleared a path for drug-resistant strains. Now scientists are worried about an additional selective force on bacteria: the rising tide of antibacterial products. In the last decade, more than 1,000 new products have flooded the market: hand soaps, lotions, detergents, toothbrushes, toys, cutting boards, kitty litter, socks, bicycle pants, pillows, highchair trays, dog beds, potty seats, shopping carts, fabric cleaners, kitchen counters, even ballpoint pens. In some grocery stores, searching for a cleanser free of antibacterials is like trying to find an eight-track tape of Rod Stewart. A survey released in 2000 by Eli Perencevich, an infectious disease specialist at Boston's Beth Israel Deaconess Hospital, found that 76 percent of liquid soaps and 29 percent of bar soaps—45 percent of soaps overall—contained antibacterials. Ironically, antibacterial-containing products are no more effective in cleaning away germs than plain soap and water. And they may actually introduce their own risks: getting rid of good bacteria while fostering drug resistance in disease-causing bacteria.

What's wrong with all this cheerful marketing of a germ-free world? "Life is not sterile," explains Vincenza Snow, an internist in Philadelphia. "There are actually good bacteria that live on our skin and in our mucosal lining that protect against bad bacteria. It's not a good idea to sterilize our life or our skin." If normal bacteria become drug-resistant from everyday exposure to household "germ-fighters," such life-saving products as the antibacterial soaps used in ICUs and pediatric units could lose their effectiveness.

Until recently, scientists hadn't paid these products much attention. They assumed bacteria couldn't gain resistance to such products as the popular triclosan, a growth-inhibitor and bacteria killer that apparently kills indiscriminately, like a bomb. Industry officials chimed in that since antibacterials have been used for decades in hospitals, surely resistance would have shown up by now. But the recent flood of antibacterial products exceeds anything from the past. In 1999, investigators at St. Jude Children's Research Hospital in Memphis showed that triclosan interferes with a specific biochemical process inside bacterial cells—and that through genetic mutation, bacteria can find a way to resist. In fact, they already do: triclosan-resistant E. coli pump the chemical out of the cell, the same efflux mechanism they use to squirt out tetracycline.

A hyperclean childhood environment may set us up for lifelong medical problems. According to the so-called hygiene hypothesis, early exposure to certain pathogens may actually be healthy. "The maturation of the immune system requires a certain amount of contact with bacteria and maybe viruses in order to be put straight," says Stuart Levy. Recent studies demonstrate that a larger proportion of people raised in highly hygienic environments suffer allergies and asthma. A 2000 report found that day care attendance during infancy (even if it increases the risk of drug-resistant infections) or the presence of older siblings in the home protected older children against asthma. "As we've added one more dimension to hygiene—use of antibacterials in the home—where will the infants have contact with the normal bacteria?" Levy asks. "My concern is that we'll see a lot more allergy, but we won't see it until they're age nine, ten, eleven, twelve—and by then, you can't go back, it's done. If you don't need these products, why take the chance?"

Secrets of Success

What is it about bacteria that make them preeminent survivors? What attributes do they possess that stir scientists to breathless professions of awe?

"Never underestimate an adversary that has a three-point-five-billion-year head start," says Abigail Salyers, a microbiologist at the University of Illinois at Urbana–Champaign. A bacterium's sole purpose is to duplicate itself. A single bacterial survivor can found a teeming colony. Above all, bacteria swap genes more easily than any other living creatures, genes that are often bundled into block grants of trouble. Even the most distantly related species can exchange DNA.

To understand the genius of bacterial defenses, you first need to know a bit about how antibiotics themselves work. Most of the 160 or so around today are variations on 16 basic chemical compounds. Some—macrolides, tetracycline, and aminoglycosides—block the manufacture of proteins needed to make new cells. Others—penicillin, cephalosporins, and glycopeptides—gum up the synthesis of a bacterium's protective wall, so that the wall falls apart and the cell dies.

Yet others—fluoroquinolones, trimethoprim, and sulfamethoxazole—interfere with the DNA replication needed to live and multiply.

In response, bacteria have learned to hogtie drugs. They can shut the gates, camouflage themselves, spit out the drugs, waylay them, change them, or just plain destroy them. The simplest way to thwart antibiotics is to block them from entering the cell. Tetracycline-resistant bacteria pump the antibiotic out faster than it accumulates inside, a ploy known as a drug efflux mechanism (and aped by cancer cells, fungi, and parasites to bail their own chemotherapeutic nemeses). Other bacteria make enzymes that target and destroy antibiotics like guided missiles: penicillinases destroy penicillins, cephalosporinases wipe out cephalosporins. Enzymes can also hobble antibiotics such as streptomycin and other aminoglycosides by slapping them with a chemical residue. Pneumococci actually rebuild the proteins that penicillin targets, the so-called penicillin-binding proteins in the cell wall. Bacteria resistant to trimethoprim and sulfonamides make a phony copy of the enzymes those drugs normally act on, a diversionary tactic that leaves the bacterial cell safe. Staph bacteria resistant to vancomycin construct a wall that acts like a sponge, absorbing the antibiotic before it can penetrate the cell.

VRE takes the cake. It completely changes the ingredients it uses to make its cell wall, ingredients that are normally targeted by vancomycin. Switching to a brand-new biochemical recipe—a recipe biologists have never seen before—it can make a perfectly functional barrier. "That's a real *tour de force*," says David Hooper, an infection control director at Massachusetts General Hospital. "What that tells me is: No matter what we come up with, over time bugs are going to figure out a way to get around it."

Most of the microorganisms that have given us antibiotics come from soil. The traditional explanation is that these organisms stage some kind of germ warfare with each other, protecting their turf against the intrusion of other species. Recently, scientists have mulled other theories. Maybe bacteria and fungi make antibiotics in the vanishingly small amounts they do, not to kill off their enemies, but to signal and interact with each other. Maybe antibiotic resistance genes, which have been around eons before the pharmaceutical industry, originally had more mundane housekeeping tasks in the cell, and just happened to be ready-made for drug resistance when the drugs poured on.

Odd as it may sound, it's the microorganisms that *produce* antibiotics that are also the source of resistance genes to that antibiotic. The cluster of genes that VRE uses to fend off vancomycin, for instance, is the same cluster found in the soil organism that makes vancomycin, *Amycoloptosis orientalis.* It is likely that bacteria need these resistance genes to help them fend off their own toxic products. The practical implication is that whenever scientists stumble on a new antibiotic, they should be able to find in the organism that churns it out genes that other bacteria will deploy to resist the drug—a preview, in other words, of how resistance will arise in humans.

Bacteria live in a tariff-free zone. Resistance genes move freely across species—even between Gram-positives and Gram-negatives, a distance as vast in phylogenetic terms as that separating an amoeba and Albert Einstein. *Staphylococcus* shares resistance genes with *Listeria. E. coli* has traded DNA with *Pseudomonas* and *Neisseria.* MRSA became the beast that it is by steadily building its genetic stockpile. "The exchange of genes is so pervasive," writes Stuart Levy, "that the entire bacterial world can be thought of as one huge multicellular organism." This fluidity may enliven the social lives of these austere one-celled organisms, but it's a disaster for public health. The history of antibiotic resistance reads like a genetic travelogue.

How *do* they do what they do? Let's start with genetic mutation, the basic engine of evolution. In duplicating itself, a bacterium may accidentally alter a gene. That change may give it new armor against certain drugs, and a competitive advantage over its neighbors that lack the new gene. A single mutation helped the tuberculosis bacterium resist streptomycin; in the early 1990s, a sequence of other mutations lent it resistance to other drugs. The newfound strains of linezolid-resistant staph and enterococcus both rely on chromosomal mutations—which can be passed on to progeny—to evade the drug. But it can take many throws of the genetic dice before an organism mutates into resistance. Pneumococcus, for instance, took a quarter century to resist penicillin, and group A *Streptococcus* hasn't even managed that.

Some genetic trade routes link two closely related species. In a process known as transduction, a bacteriophage virus infects a bacterium, incorporates that cell's resistance genes into its own genome, then copies itself and bursts out to infect new bacterial cells, carrying along the resistance genes. Some scientists think *Staphylococcus aureus* acquired its methicillin resistance gene this way. In a genetic exchange program called transformation, a bacterium scavenges floating pieces of DNA released by a dead bacterial cell from a closely related species, integrating the "naked" DNA into its chromosome. Through transformation, *Pneumococcus* first became penicillin resistant.

The best genetic get-rich-quick scheme, however, is not to save or scrounge but to steal resistance genes right off the shelf from a donor cell. "If bacteria had a brain," says Abigail Salyers, "this would be a no-brainer."

The particulars of "horizontal gene transfer," as it is called, first came to light in the 1950s. In 1959, Japanese scientists isolated a *Shigella dysenteriae* that resisted four different kinds of antibiotics: tetracycline, sulfonamide, streptomycin, and chloramphenicol. Then as now, *Shigella* was a major cause of dysentery in Asia and Central and South America. The pokey mutability of the bacterium's own chromosome couldn't account for this sudden wealth of resistance. In fact, Japanese scientists calculated that the microbe would have had to divide 10 million × 10 million × 10 million × 10 million times—far longer than bacteria have been on earth, or than there's been an earth—to have pulled off this quadruple hat trick. At the same time, researchers found strains of benign *E. coli* resistant to the same four drugs. In what direction the gene package traveled—from the *E. coli* to the *Shigella* or the other way around—is not known. What's important is how they rode: on circles of DNA known as plasmids which, carrying genes resistant to several drugs, enabled the consignees to quash agents they'd never seen before.

"It was an incredibly novel aspect of science," says Julian Davies, professor emeritus of microbiology at the University of British Columbia. So novel that the first papers published in Japan were rejected in the United States, and at least one researcher lost his job over the new proposition. According to Davies, "The convention was that genes just don't move around. If genes move around, how could you possibly have species? But genes do move around, and one wonders what 'species' really means."

Plasmids are mobile loops of DNA separate from the bacterium's chromosome. You can think of them as shopping carts where bacteria store genes that they only need now and then. Plasmids pack genes for all kinds of emergencies: to withstand temperature extremes and ultraviolet light, to survive a torrent of gastrointestinal juices, to degrade environmental pollutants—and to resist antibiotics. Plasmids reproduce and migrate during the tryst known as conjugation, in which two, often unrelated, bacterial cells briefly draw together and pass their genes to the other in several ways. Antibiotics such as tetracycline may act as aphrodisiacs, prompting bacteria to mate with abandon and indulge in an orgy of gene transfer. Over time, plasmids and their bacterial hosts can enter a symbiotic relationship, in which the growth of the host depends on the plasmid—one reason that the drug resistance bestowed this way is hard to reverse.

Resistance genes can also ride on mobile pieces of DNA called transposons. Known colloquially as "jumping genes," transposons, unlike plasmids, are truly free spirits. They don't need the host cell to survive; they can dash off whenever they like. A special kind of transposon known as a conjugative transposon plays a major role in antibiotic resistance. Conjugative transposons bearing resistance genes can jump from their seat on the donor chromosome onto a plasmid, then during mating stow away on that plasmid to another bacterium. Once they're settled again, they can jump from the plasmid to the new host's chromosome—a stable perch from which to do their job.

This genetic free-for-all creates problems for medicine. At Emory University, population geneticist Bruce Levin has found that resistance to an antibiotic persists long after doctors have stopped using that drug. This is a blow to researchers who hoped that bacteria resistant to overused drugs would "evolve backward" if spared the drug. Analyzing the fragrant contents of diapers from a daycare center, Levin found that a quarter of the *E. coli* lurking between the folds resisted streptomycin, a drug rarely used in the past 30 years. Although in evolutionary theory resistant bacteria are presumed to be more genetically weighed down and therefore less fit to compete, Levin suspects that after *E. coli* gains drug resistance, it evolves a second compensatory mutation that keeps it from backsliding to a state of drug sensitivity.

"Resistance genes," says Abigail Salyers, "are easy to get and hard to lose." The mechanics of plasmid transfer leads to an exponential rise in resistance genes. And a single antibiotic to treat an infection can provoke resistance to other drugs. Take tetracycline for a chlamydia infection, say, and your gut bacteria can suddenly become resistant to half a dozen different, structurally unrelated antibiotics. One reason may be a master switch—dubbed MAR, for Multiple Antibiotic Resistance—on the cell's chromosome. "It's almost as if bacteria strategically anticipate the confrontation of other drugs when they resist one," writes Stuart Levy, whose lab discovered MAR in Gram-negative organisms such as *E. coli* and *Salmonella*. Moreover, though resistant strains can drop in number if they lose out in competition with drug-sensitive strains, they seldom disappear completely. That means there's always a residue of resistant bacteria around, ready to multiply if the right antibiotic rains down on them.

Resisting Resistance

At the start of 2001, the U.S. Department of Health and Human Services announced an "action plan" to combat antibiotic resistance. But in the United States, it's impossible to get a firm grasp on the threat. The National Nosocomial Infection Surveillance system, a CDC-sponsored voluntary reporting effort among U.S. hospitals, is not comprehensive. Indeed, no federal agency tracks all resistant infections or adds up their human and financial costs. And admittedly, statistics can play tricks. If raw data were all that counted, large teaching hospitals with very sick patients would look like they're doing a bad job—which is why you'll probably never find out about a hospital's antibiotic resistance problems before you or someone you love is admitted. Even then, it's hard to get the lowdown.

Fighting antibiotic resistance is like wrestling Kali, the many-armed

Hindu goddess of life and death. It is not one problem but many. While some observers offer visionary solutions, and others practical nostrums, nearly everyone involved complains that deeper institutional battles are slowing progress. On one fact everyone agrees: the propensity of bacteria to alter themselves in the face of threat will never be halted, merely contained. The most that public health experts hope for is to transform antibiotic resistance from a crisis—as it is now—to a routine annoyance.

One way is to speed up the pharmaceutical production line. Today's antibiotics belong to a surprisingly small number of structural groups, most descendants of older progenitors harvested from nature. A recent exception is linezolid, a synthetic drug and the first new antibiotic in 35 years. Because the compound is manmade—and would theoretically not encounter any preexisting resistance genes in nature—researchers expected that the drug would have a long shelf life. But dismayingly, as Mary Jane Ferraro and others have discovered, linezolid-resistant forms of both staph and enterococcus appeared almost instantly. Tomorrow's antibiotics may be plucked from huge collections of synthetic molecules which, with new technology, can be tested in huge numbers against resistant organisms. Gene sequencing will accelerate this process by laying out complete genetic sequences of bacteria—and therefore, new biochemical pathways to target.

A second approach is to take a well-respected antibiotic that has lost its clout and give it a second life. Stuart Levy's lab, for example, is bioengineering new tetracyclines that can block the pump in bacteria that spit it out.

A third, somewhat theoretical, approach deploys bacteriophages—the viruses that target bacteria. The inspiration for their therapeutic potential goes back to a 1917 report by Felix d'Hérelle, a young French bacteriologist, who observed that the bloody stools of soldiers struck by dysentery contained invisible agents, thought to be viruses, that could destroy the offending bacteria. D'Hérelle went on to predict that these viruses, or phages, could cure all manner of bacterial disease. Since a phage virus makes copies of itself—a single phage produces more than a billion progeny in three to five hours—it is potentially the only drug that could multiply itself in the body. Since the 1930s, phage work proceeded apace in the (now former) Soviet Union, with claims of success against *Enterococcus* and other mostly enteric and skin pathogens. More recently, researchers have completed promising experiments with mice and are embarking on human trials. Today, several American biotech companies have taken the baton, hoping to cultivate phages against VRE, MRSA, and other bugs.

Drawing on the fact that no infectious diseases have ever been eliminated by antibiotics, other experts are focusing on vaccines as the ultimate end-run around resistance. During the 1980s, *Haemophilus influenzae* type B (HiB) was the leading cause of meningitis in American infants and a major agent of blood poisoning and pneumonia. But since 1989, when the HiB vaccine came out, HiB meningitis has virtually disappeared in the United States. A pneumococcal vaccine approved for children in 2000 has likewise proven effective against serious complications such as bloodstream infections and pneumonia, and is expected to curtail not only kids' ear infections but the resulting overuse of antibiotics. More challenging will be making vaccines against staph or enterococci, since these organisms are benign lodgers against which the body doesn't mount a strong immune response—making them fuzzy targets for vaccines.

There are also ecologic approaches to defusing resistance. Researchers talk of biodegradable antibiotics that would lose their resistance-breeding potency once released into the environment. Another proposal would harness the evolutionary forces that foster resistance and turn them to our favor. Picture a bacterial reclamation project, in which resistant flora are replaced with friendly—i.e., drug-sensitive—microbial species. In Stuart Levy's mind, that requires first stopping the use of certain antibiotics so that the selective pressure is off, and then literally restocking the environment with susceptible bacteria—say, by spraying drug-sensitive enterococci in intensive care units. Other scientists, however, worry that these newly added bacteria will also become resistant.

***

But technological fixes take time. At the CDC, researchers estimate that a third of the 150 million annual outpatient prescriptions for antibiotics are unnecessary, either because the illness is not bacterial or because the bacteria aren't sensitive to the antibiotics. According to a 1998 study, antibiotics were prescribed 66 percent of the time during office visits for coughs, even though most coughs—and sore throats and runny noses—are caused by viruses, which antibiotics can't touch.

In response, the American College of Physicians in 2001 issued new recommendations urging doctors to use antibiotics for common respiratory illnesses only when they strongly suspect a bacterial cause. Respiratory infections account for three-quarters of all antibiotic prescriptions. "We were all taught in medical school that anybody who was coughing up something green had a bacterial infection. But now we know that this is just the natural evolution of a viral infection," says internist Vincenza Snow, who coauthored the ACP guidelines. "As physicians, we have to unlearn some of the things we've been taught. And we have to reeducate our patients." In 2000, the federal Agency for Healthcare Research and Quality found that the vast majority of children with middle ear infections recover without antibiotics—so waiting for a few days won't put kids at risk, even when they require treatment later for their illness. Patients who do take antibiotic drugs are being reminded to take all their pills, so as not to promote resistant bacteria.

Doctors and patients may finally be getting the message. A CDC survey found that between 1989 and 1998, antibiotic prescriptions for children under 15 dropped by 30 percent. Unpublished data show that, among adults, antibiotic prescriptions fell by 10 to 15 percent between 1995 and 1999. But according to experts, these declines must double before the upward trend in antibiotic resistance reverses.

To curtail antibiotics, doctors need to know precisely what's ailing their patients. That's not always easy. Today, there are no tests to instantly pin down what bug is causing an infection and what drugs it resists. Besides, managed care groups are trying to save money by cutting back on tests, including microbiologic cultures, because it's cheaper to prescribe broad-spectrum drugs and wait to see if patients get better. If new gene-based tests were around tomorrow, would doctors use them? More to the point, would health maintenance organizations pay for them?

Meanwhile, the Food and Drug Administration has focused on cutting agricultural antibiotics crucial in human medicine—playing a long-delayed game of catch-up. In 1977, when the agency first proposed removing penicillin and tetracycline from growth promoters, Congress blocked the move. "That set back the national agenda for 20 years," says the CDC's Fred Angulo. By 2005 or so, the agency intends to establish regulations that will automatically remove an antibiotic from farm use if it's proven to cause high levels of drug-resistant infections in people. The focus will be on preserving drugs crucial for human medical therapy. For their part, consumers may think twice about buying household products that trumpet their "germ-fighting" prowess.

Many public health experts are betting on an up-to-the-minute information system that could tell doctors what resistant organisms are in their community and therefore what drugs to steer away from. But there are times when all the information in the world won't help a doctor make a tough call—times, in fact, when a doctor's drive to deliver the best clinical care seems at odds with public health campaigns against resistance. Doctors, after all, are advocates for *their* patients— not for *all* patients, not for global ecology. "It gets into some extremely difficult philosophical questions," says the University of Maryland's Glenn Morris. "I'm the bad guy. I wear the black hat. Because I tell doctors they can't use drugs that they think they need for their patients." The tension between what's right for the individual and what's right for public health becomes especially poignant at the end of life. Should a doctor give last-resort antibiotics to a patient who is sure to die soon anyway—even if it means raising the risk of drug resistance in another patient five feet away?

"When you're looking a patient in the eye, or when you're looking a farmer in the eye who's got a flock of sick chickens, it's very difficult to say, 'I'm not sure you need that antibiotic,'" says Morris. "If it's your grandmother sitting in the hospital with a bad infection, would you prefer that the doctor gave her three antibiotics to make sure he's covered whatever's causing the infection, or that he said, 'Let's just give her one and wait for the culture to come back'? The answer is, of course, 'Do everything you can, Doc!' Multiply that by hundreds of thousands of cases, and you have overuse in hospitals."

The daunting challenge of antibiotic resistance owes itself to institutional complacency as well. Few researchers today have been trained in the basic biology of bacteria—because few university laboratories are left to specialize in the problem of drug resistance. Discouraged by the dearth of funding for the field, young investigators have chosen other disciplines.

Nor has the pharmaceutical industry been keen to tackle drug resistance. The biological paradox that antibiotics sow the seeds of their own failure is matched by an economic paradox: if a new antibiotic is good, regulators will ask doctors to use it sparingly and the drug won't quickly recoup its investment. Despite the development of linezolid-resistant staph and enterococcus, will its manufacturer push the new drug? "The pneumonia market is many orders of magnitude larger than the enterococcus market," says John Quinn, a physician at the University of Illinois at Chicago, who first published evidence of linezolid-resistant VRE. "And if you're sales manager for a new agent, and your compensation is linked to the number of doses of that drug being administered, it's got to be awfully tempting to promote the use of the drug in respiratory infections. I'm not making a specific accusation that the company is doing that. But certainly the history of antibiotics is one of enthusiastic use followed by overuse followed by resistance."

Making matters worse, most of the easy drug targets have already been discovered, so that it's even more costly to find chinks in bacterial armor. This dilemma could discourage drug companies from investing in what typically takes at least seven years and costs $350 million to bring a new antibiotic to market. At the end of that road, what companies want is a blockbuster like Viagra or Prozac, something that will stand out from its competitors and make an immediate $2 billion to $4 billion. "In most companies, antibacterial discovery and development is a Cinderella field," says David Shlaes, vice president of infectious disease research at Wyeth-Ayerst Research. "In the 1980s, drug companies started getting out of this business in droves, and that hasn't stopped. It remains for the pharmaceutical companies a tight market and a competitive market." To offset the economic losses of a rationed drug, companies might need to be compensated with longer patents, fast-track approval, or other marketing incentives for drug development, as orphan drugs for rare diseases now receive. "When we find a drug effective against VRE," says Glenn Morris, "we should use it only against VRE, even if we're only talking ten thousand to twenty thousand cases a year."

Controlling hospital infections would also help cut down drug resistance. In the last two decades, the rate of hospital-acquired infections has risen in part because today's patients are sicker and more vulnerable. Yet there are few incentives for hospitals to monitor infection rates. A head-in-the-sand attitude prevails in this country: If hospitals don't look for resistant infections, they won't find them and they won't have to fret about them. Besides, the thinking goes, why should one institution pay a public relations price for a problem entrenched in every hospital? In our competitive and cost-cutting times, infection control departments are perceived as nests of unbillable activities. "If I've got to make a choice between adding an infection control nurse or a hospital epidemiologist versus a cardiac surgeon," explains the CDC's William Jarvis, "I'm choosing the cardiac surgeon, who can bring me several million dollars a week."

Yet a mathematical model developed by Emory University's Bruce Levin and Harvard School of Public Health's Marc Lipsitch underscores the need for infection control. It says that the influx of fresh, drug-sensitive bacteria carried by newly admitted patients should be able to outcompete resistant strains in hospitals—as long as those hospitals vigorously try to prevent antibiotic resistance. Judging from the few documented interventions attempted in this country—controlling VRE rates by segregating colonized patients, reducing resistant bloodstream infections among cancer patients through infection control, and others—prevention can save money in the long run. But that will happen only if every hospital and doctor shoulders the task equally.

Hospital administrators claim that outside pressures force doctors to err on the side of prescribing antibiotics, even when unnecessary. Four floors above the emergency room of Boston's Beth Israel Deaconess Medical Center, ambulance sirens repeatedly pierce the air. From that perch, Robert Moellering Jr., chair of the Department of Medicine, must grapple with the legal repercussions of pressured medical decisions. "We're in this position of trying to get physicians not to overuse antibiotics," he says. "But I have never yet seen a physician sued for giving antibiotics inappropriately. I've seen all kinds of them sued for *not* giving antibiotics." This fear of malpractice suits, Moellering says, also stokes antibiotic resistance.

We're not yet on the brink of a post-antibiotic age. Compared to the last turn of the century, we have better hygiene, sanitation, and nutrition, safeguards that have always prevented disease from starting and spreading. And even though 70 percent of the bugs that cause hospital infections defy at least one antibiotic, there are usually four or five others to try.

But like love, drug resistance is here to stay. "We have squandered an immense resource, much of it a true natural resource, by using it frivolously, inappropriately, and ineptly. . . . Above all, we polluted our hospitals, and in doing so we set up a truly global laboratory: a colossal Darwinian experiment in enhanced evolution," writes the physician Imre J. P. Loefler, in the *Lancet*. "We wasted resources that, if husbanded, would have helped us treat infections perhaps for hundreds of years. We squandered because of ignorance, consumerism, mercantilism, cynicism, and carelessness."

Tracking the relentless rise of resistance at the University of Maryland Medical Center, which had the dubious honor of being one of the first to report VRE in this country, Glenn Morris feels a sense of dread. "This is the gradual, inexorable movement of a glacier headed directly toward your town. It's slow, silent, unexciting. But every year you look at the 'Percent Resistant' isolates in your hospital. At times it feels like we're hanging on by our fingernails."

Chapter 5

The Once and Future Pandemic

Johan Hultin didn't own a decent set of autopsy tools. It had been eight years since he'd retired, and in that time he'd had no occasion to excavate a human body for clues to its downfall. He was adept at improvising, however. Before leaving San Francisco for Alaska, he'd appropriated his wife's pruning shears (that wasn't much of a stretch; pathologists prefer garden-variety pruning shears for opening the rib cage) and, from the kitchen, a broad carving knife.

It was the middle of August 1997. On this day, like every day, the wind blew hard off the wide bay surrounding Brevig Mission, 65 miles northwest of Nome and just south of the Bering Strait. Wielding shovel and pickax, Hultin dug toward permafrost, where 72 Eskimos (as Caucasians called them) had been buried, victims of the 1918 influenza pandemic, which claimed more lives than any other infectious outbreak in human history. Looking east as he excavated, Hultin saw bare tundra tussock leading to a low range of coastal mountains. Looking south, a red shed near a gravel airstrip. To the west, a tall bluff, broad beach, and gray wind-whipped water. To the north, the village cemetery where weather-beaten wooden crosses marked tombs of the native Inuit and granite stones decorated graves—some going back to the turn of the century—of the village's Lutheran missionaries.

Hultin was here on a rescue mission of sorts, an expedition to pluck from permafrost a long-dead virus whose secrets might save millions of lives in the future. As far back as graduate school, half a century earlier, he had dreamed of cracking the secrets of the 1918 flu. A native of Sweden, Hultin had traveled to the United States in 1949 for a six-month course in microbiology at the State University of Iowa, an institution then famed for its influenza research. In 1951, fired up by a teacher's offhand remark that live virus may yet reside in permafrost-preserved victims, Hultin had persuaded two professors and a paleontologist friend in Fairbanks to set out in search of the pathogen. Hopping rides with bush pilots, he assessed three promising sites described by missionary records to make sure the local permafrost level hadn't shifted since the burials 33 years earlier. Only in Brevig Mission—an isolated village where the native population was just beginning to relinquish millennia-old customs of subsistence hunting and fishing—were the burials intact.

As the 1951 excavation began, the villagers were asked to stay away, in case particles of live virus should escape, potentially triggering a new pandemic. In the bleak, windy landscape, Hultin and his crew wore surgical masks and gloves. To make sure the specimens didn't get contaminated, they built a fire in a gasoline-powered camp stove to boil water for sterilizing instruments. Ultimately, they found four bodies with soft tissue. Back in their Iowa City lab, Hultin and his colleagues tried everything they knew to revive the virus (all the while following the same laboratory precautions scientists were using to protect themselves against the deadly bacillus that causes tularemia, an agent being perfected at the time for biowarfare). They inoculated the virus into hundreds and hundreds of chicken eggs. They injected it into mice and rats. They squirted it into the nostrils of ferrets, animals in which human flu viruses were first isolated.

Nothing worked. The virus was dead, irretrievably so. For 46 years, Hultin's dream would remain on—and in—ice.

***

In the 1918 influenza pandemic, Brevig Mission, like most Eskimo outposts, had been decimated. On November 15, the first villagers had begun dying. Within five days, 90 percent of the village—72 of 80 residents—had perished. In one nearby Eskimo outpost, York, not one soul survived. As historian Alfred W. Crosby wrote in *The Forgotten Pandemic,* "Spanish influenza did to Nome and the Seward Peninsula what the Black Death did to fourteenth-century Europe."

The flu had probably found its way to Brevig Mission (then known as Teller Mission) on the breath of unsuspecting travelers: passengers on a supply ship to Nome, then the men who brought those supplies to the nearby trading post of Teller, then Eskimos from the mission who loaded their dogsleds with supplies there. As the sickness spread, a pall descended on the gloomy outpost, where late-autumn sun lasted only four hours a day. "The sick were constantly moaning and groaning," wrote one survivor, Clara Fosso, the Lutheran minister's wife. "Outside, the loose wild dogs howled like wolves."

A party from Teller traveled 14 miles by dogsled to offer whatever assistance they could. They shot the prowling dogs and searched for signs of life in the igloos. One housed 25 dead bodies. Another contained a pile of human bones—leftovers of a canine meal. The men pierced the seal-gut window of another abode to peer inside at a group of corpses. "Much snow had drifted in," Fosso wrote. "Luckily, one thought he saw something move in the corner of the igloo. As they shouted down, three frightened children popped from under the deer skins screaming. They virtually had to be captured for they seemed to be in a wild stupor."

Officials at the U.S. Army base at nearby Fort Davis brought in gold miners from Nome to dig a collective grave. Using steam generators, the miners melted a long rectangular gash in the earth. The victims were each tied with a rope around the chest, dragged across the ice, and laid side by side at an army regulation depth of six feet. Two tall wooden crosses, visible atop the bluff from the sea, marked the grave.

That year a Christmas celebration was conducted in the Eskimo language. The mission church was converted into an orphanage. Slowly, the village recovered.

The 1918 Flu

The most dangerous misconception about flu is that it is nothing more than a bad cold. The truth is, flu is the illness many public health experts dread most, because it is the only disease that results in what epidemiologists call "excess mortality." The seasonal spike it brings in the death rate comes not only from influenza and pneumonia, but also from more cardiac failure and pulmonary disease and diabetic complications than in flu-free times. No other illness—not AIDS, not Ebola, not TB—causes such statistical skewing, such profound disturbances across populations. And by constantly changing, the flu virus mocks our annual attempts to protect ourselves with vaccination. It may be the cleverest, most agile microorganism known.

A flu epidemic is always bad news. It can also turn into a globe-girdling disaster. Since the sixteenth century, there have been 13 flu pandemics: sudden worldwide epidemics against which no human possessed even partial immunity. The 1918 flu was the most fearsome, an unprecedented panic of which the nightmare in Brevig Mission was merely a miniature. Experts say it's a matter of time before another flu pandemic, perhaps just as deadly, erupts.

As generals launched the final offensives of World War I, this monster variant of the influenza virus was quietly unfolding. Between 20 million and 40 million people died worldwide (some estimates run to 100 million, or 1 percent of the population at that time), at least one-fifth of the population suffered aches and fever, and most of those not afflicted were symptomlessly infected. The new strain of a seemingly familiar disease killed twice as many people as those who died from battle wounds over the war's four horrific years. Only the remote island of St. Helena in the South Atlantic and a handful of Pacific islands were spared. In 1918 and 1919, at least 500,000 Americans, of a total population of 103 million, died—a figure that included 43,000 soldiers. India lost at least 12.5 million, or 4 percent of its population.

In terms of sheer numbers, if not percentages, nothing in history approached this galloping mortality. The Plague of Justinian, thought to be bubonic plague, which began in AD 542, killed an estimated 100 million people, but took 50 years to do so. Over four centuries, the Old World pathogens of smallpox and measles reduced the New World population by perhaps 90 million. The 1918 flu took most of its victims in four months.

The virus mounted an appalling blitz. It sometimes killed within 48 hours, striking down victims with high fever, delirium, insurmountable pain, and weakness in the midst of daily routines. People descended from apparently robust health to death's door in an hour. A man would board a streetcar for work and fall dead before he'd reached his stop. Pedestrians halted on the sidewalk, clutching at lampposts, then slid to the pavement unconscious. Pregnant women suffered frightening rates of miscarriage and premature labor, and among those flu victims who prematurely lost their babies, 41 percent died.

"They're as blue as huckleberries and spitting blood," one New York City physician told a colleague. Cyanosis, caused by oxygen deprivation, was a common sign of imminent death. A young doctor at Fort Devens, Massachusetts, wrote: "Two hours after admission they have the mahogany spots over the cheek bones, and a few hours later you can begin to see the cyanosis extended from the ears and spreading all over the face, until it is hard to distinguish the colored man from the white." So swift was the viral attack on the immune system that many victims died not of secondary bacterial pneumonia, which usually deals the final blow, but of the massive hemorrhaging caused by the virus itself. Among those patients who did go on to develop pneumonia, 35 to 45 percent died.

The new virus had made its formal debut on March 4, 1918, when soldiers at Camp Funston, Kansas, began flocking to the base hospital complaining of fever, headache, and backache. Within a month the virus had reached most American cities. American doughboys, jammed into troopships, ferried the germ to Europe. Yet while it clearly represented a new round of flu, this spring wave didn't set off alarms. In June, the Spanish wire service cabled Reuters in London: "A STRANGE FORM OF DISEASE OF EPIDEMIC CHARACTER HAS APPEARED IN MADRID. . . . THE EPIDEMIC IS OF A MILD NATURE, NO DEATHS HAVING BEEN REPORTED." In so informing a world at war, Spain—a nonbelligerent with no press censorship—inadvertently bestowed its name on the novel infection. By July the Midwest émigré with a faux Mediterranean pedigree had spread around the world.

Summer saw an apparent viral retreat. The number of new cases dwindled, though the proportion of serious illnesses rose ominously. Then, in late August, a deadly variation exploded simultaneously in three far-flung locales: the French port city of Brest (a major disembarkation point of American soldiers), Boston (where U.S. troops returned from the battlefield), and Freetown, Sierra Leone (where the British Navy vessel HMS *Mantua* docked).

In late September a U.S. Public Health Service report registered an unmistakable note of worry: "The disease now occurring in this country and called 'Spanish Influenza' resembles a very contagious kind of 'cold' accompanied by fever, pains in the head, eyes, ears, back or other parts of the body, and a feeling of severe sickness. In most of the cases the symptoms disappear after three or four days, the patient then rapidly recovering; some of the patients, however, develop pneumonia, or inflammation of the ear, or meningitis, and many of these complicated cases die."

The microbe's target was not only the lining of the main branches of the respiratory tract, where flu usually takes up residence, but also the tiny air sacs at the terminal branches of the respiratory system, deep in the lung—the alveoli, where oxygen and carbon dioxide are exchanged. Composed of hundreds of millions of these delicate air-filled structures, healthy lungs are light and buoyant, able to float in water. Lungs removed from victims of the 1918 flu were hideously transformed: dense, heavy, the alveoli saturated with bloody fluid.

Patients died by drowning. As rigor mortis set in, bloody froth streamed from their noses, staining the bedsheets.

Just as shocking as the virus's deadliness was its choice of victim. Most flu epidemics disproportionately claim the very young and the very old. If a typical flu pandemic is traced on a graph—with the horizontal x axis representing the age of victims, and the vertical y axis representing death rates—the resulting age-specific mortality curve is a U shape. The 1918 flu singled out people in the prime of life. Its graphic representation is a rough W, with an eerie peak in the middle. All over the world, about half the deaths were in the 20 to 40 age group.

Though the jury is still out on the true birthplace of the 1918 flu, the story of the virus's midwestern provenance, burnished by retelling, has become legend. That fall, farmers noted that just about the time people were becoming sick, pigs also started to suffer from the disease—for the first time in history. For many years the question was: Did pigs give the flu to people, or the other way around? But the American cornbelt may simply have been where the disease was first *noticed*. In September 1918 the U.S. Surgeon General's report on the Spanish flu alluded to an Asian origin. "Some writers who have studied the question believe that the epidemic came from the Orient and they call attention to the fact that the Germans mention the disease as occurring along the eastern front in the summer and fall of 1917." British virologist John Oxford has found reports of highly lethal respiratory infections with cyanosis in French army camps during the winter of 1916–1917. Nevertheless, a decade after the Spanish flu fizzled out, epidemiologist Clifford Gill wrote in *The Genesis of Epidemics:* "[A]ll authorities are agreed that pandemics of influenza can almost invariably be traced to the 'silent spaces' of Asia, Siberia, and Western China." Today, Hong Kong–based virologist Kennedy F. Shortridge, who has studied influenza for nearly 30 years, calls southern China an influenza "virus soup." Something about China—perhaps its timeworn agricultural practices, its dense mingling of humans and livestock and birds—makes it a cauldron of this deadly infection.

The Science of Influenza

The influenza virus is "a wily adversary," says virologist Robert G. Webster, who has studied the virus's ecological niches for nearly 40 years. "Just when you start to think that you've understood what it can do, then it pulls another one out of the bag. It's one of the most crafty of all infectious disease agents. It's got such a repertoire of tricks."

Since its discovery 70 years ago, the virus has drawn the interest of a small but dedicated cadre of scientists. Some, like Webster, shuttle between the outdoors and the lab, some toil at the bench with tissue cultures or molecular technologies, others focus on bedside and population data, and yet others devise institutional protocols for dealing with the next pandemic.

"Influenza" derives from the Italian *influentia*—to influence, reflecting the medieval belief that astral or meteorological forces were behind the sudden and unexpected appearance of epidemic disease. By 1918 scientists knew that they were dealing not with a cosmic mechanism but a microscopic one. In the wake of the germ theory and the decline of such scourges as cholera, tuberculosis, diphtheria, gonorrhea, typhoid, and scarlet fever, they assumed that the pandemic swirling around them would likewise offer up a well-defined bacterial target. The odds-on candidate was Pfeiffer's bacillus, which frequently proliferated in the throats of patients with the Spanish flu. (Today, we call the bacterium *Haemophilus influenzae*—a misnomer, since it does not cause flu but rather bacterial meningitis.) But the oddsmakers were wrong. The flu virus, then invisible and as always constantly mutating, would prove more fugitive than the quarry of Pasteur and Koch. As the *Journal of the American Medical Association* observed in October 1918: "The 'influence' in influenza is still veiled in mystery."

By then, researchers had known for two decades that some diseases are caused by pathogens smaller than bacteria. These came to be known as "filtrable" viruses, because they could slip through the superfine porcelain filters that trapped bacteria. Now we simply call them viruses. When scientists tried injecting filtered influenza viruses into lab animals and humans (in the days before sturdy informed-consent procedures), they could not produce the flu. It wasn't until the late 1920s that pathologist Richard Shope, working with the Rockefeller Institute for Comparative Pathology, published evidence of a more efficient route. He squirted filtered swine virus directly into the snouts of pigs, and finally proved—at least in that species—that the virus was the causative agent.

In 1933 research on human influenza took a giant leap. During a flu epidemic in England, Christopher Andrewes and Wilson Smith of the National Institute for Medical Research in London inoculated ferrets with washings from the throat of a convenient human flu victim: Andrewes himself. Several days later the animals were sneezing, tearing, and feverish. Eventually researchers refined the technique of transmitting the virus from ferret to ferret, and one day the circle was closed when a scientist was examining an inoculated ferret (notably squirmy creatures) and the animal sneezed in his face. A couple of days later, he came down with the flu, a groundbreaking indignity that under today's strict lab procedures probably wouldn't happen. Two years later Wilson Smith discovered that the flu virus can be cultivated in chick embryos—fertile hen's eggs—a discovery that paved the way for vaccines. In 1943 scientists finally gazed at the infinitesimal beast with an electron microscope.

The influenza virus is a member of the *Orthomyxoviridae* family, whose cousins cause mumps and measles. At about 100 nanometers in diameter, it is average in size for a virus. Under an electron microscope the virus looks like a spiky cotton ball, but the fluffy-looking sphere is actually a fatty membrane. Inside its two lipid layers are the virus's eight RNA molecules, its genetic intelligence. Studding the surface are 600 stakelike structures, like pins in a pincushion, the glycoproteins hemagglutinin (H) and neuraminidase (N), whose ever-changing antigenic raiment helps them elude detection.

Released through a sneeze or cough, the influenza virus spreads in aerosolized droplets that take at least 17 minutes to drift down from ceiling to floor and that can survive for hours on solid surfaces such as steel and plastic. Its transmissibility is legendary. In the Northern

Hemisphere, summertime epidemics on cruise ships and overland tours are increasingly part of the package, as vacationers (mostly retirees) from all over the world converge in close quarters. In a 1998 outbreak, 40,000 tourists and tourism workers caught the flu in Alaska and the Yukon Territory. Perhaps the most famous travelers' outbreak occurred in 1977, when a jet with 54 persons aboard was grounded by engine failure. Most passengers stayed on the plane during the four-and-a-half-hour delay, including a young woman who lay prostrate across two seats in the back of the cabin, coughing and feverish with flu. Within three days 72 percent of the passengers shared her pain.

Once the virus makes contact with mucous membranes in the eyes and nose, it heads directly to the epithelial cells that line the upper respiratory tract, bronchial tubes, and trachea, where it swiftly multiplies. All the while, it mutates as it divides, donning an evolving wardrobe of subtly varied—and occasionally radically different—surface attire. Within 24 to 48 hours after the virus has entered the body, it is expelled in a sneeze or cough.

A closer look illustrates the influenza virus's ingenuity. Its spiky proteins, made of folded strings of amino acids, help it latch onto the epithelial cells lining the nose and throat. Hemagglutinin, a triangular rod-shaped structure abundant on the surface of the virus, specializes in binding to a sugary molecule, sialic acid, on the surface of these cells (the name hemagglutinin was originally used because the virus was discovered to agglutinate red blood cells, i.e., make them clump together, a phenomenon that became the basis of an assay for flu viruses). The docking maneuver prompts the cell to engulf the

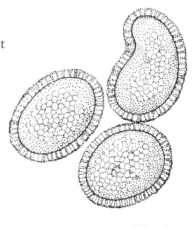

[100 nm]

Influenza virus
(family Orthomyxoviridae)
Agent of flu

virus. Soon the influenza virus—which, like all viruses, cannot reproduce on its own—hijacks the cell's machinery for its own purposes. First, its RNA and internal proteins invade the cell nucleus. There, viral proteins copy the virus's RNA strands. They also make a form called messenger RNA, which can be translated into proteins by the cell's own protein-making apparatus. Eventually, the newly made viral genes and proteins assemble themselves and bud off from the cell as fresh regiments of influenza virions. Neuraminidase, an enzyme that looks like a stalk with a mushroom-shaped head, helps free the newly minted particles from the host cell, spreading the infection.

Scientists believe that flu symptoms arise because viral growth damages colonized cells, and because the immune system, in trying to limit the damage, produces local inflammation and bodywide aches and fever. A day or so after infection, classic symptoms set in: runny or stuffed nose, dry cough, chills, fever, aches, deep fatigue, and loss of appetite. What scientists don't understand is how the virus clobbers us from head to toe, when it multiplies only in the respiratory tract and doesn't even swim around in the bloodstream.

There are two major types of influenza virus—A and B, categories that reflect differences in internal proteins. Influenza B mostly infects young children, causing mild respiratory illness. Influenza A is the stuff of which serious epidemics and deadly pandemics are made. Comprising 15 known subtypes of hemagglutinin and 9 of neuraminidase (numbered in the order they were identified), influenza A viruses attack many species of birds and mammals. Type A viruses are named according to the particular hemagglutinin and neuraminidase surface proteins they display, and are further categorized according to slight variants in their antigens—the parts of the virus that induce the immune system to make antibodies. The World Health Organization nomenclature system refers to the host of origin (for nonhuman flu viruses), the geographical site where it was first isolated, the strain number, and the year of isolation. Thus, for instance, A/Swine/Iowa/15/30 (H1N1): the classic swine strain, believed to be a close relative of the human pandemic virus of 1918.

The genius of the influenza virus is its propensity to alter itself.

Unlike a DNA virus such as the smallpox virus, whose careful genetic spell-checking minimizes mistakes when it replicates, an RNA virus uses a different polymerase enzyme to copy itself. RNA polymerase is notoriously error-prone, making every successive generation of RNA virus always at least a shade different from its predecessor. The surface proteins hemagglutinin and neuraminidase are also mutable, and because these proteins contain most of the antigenic sites that provoke, and in turn are targeted by, the immune system, an erratic guise works in their favor. Antibodies tailor-made for last year's strain of flu—and last year's antigens—cannot protect against this year's variation. Indeed, in mammals, host antibodies exert a selective pressure on the virus, giving an advantage to surface mutations that help the virus evade the immune system.

The flu virus's penchant for rapidly transforming itself in response to hostile antibodies is both an asset and a desperate survival measure. "The paradox is that this is apparently a very fragile virus," says Edwin D. Kilbourne, dean of flu research in the United States, and now a professor at New York Medical College. "Unless it mutates, and unless a new mutant is selected, it's going to disappear. It's an ephemeral virus, with remarkable tenacity and adaptability."

In general, the same subtype of human flu circulates year to year, and only one subtype circulates at a time. The H3N2 subtype, for instance, has been around since 1968. Every year, during flu season, epidemics break out because slight genetic mutations, producing small variations in the virus's surface proteins, accumulate from the year before, in a process known as antigenic drift. New combination vaccines are annually formulated to protect against the three most threatening strains of flu. But sometimes the virus's surface proteins undergo a radical substitution, and brand-new hemagglutinin and sometimes neuraminidase, which most people have never encountered before and against which they have no repertoire of antibodies, suddenly appears. With no defenses against the unfamiliar antigens, the human race gets blindsided. The result is a pandemic. While rare, pandemics sweep the world like wildfire. Three major ones broke out in the twentieth century: an H1N1 in 1918 (the misnamed "Spanish" flu), an H2N2 in

1957 (the "Asian" flu), and an H3N2 in 1968 (the "Hong Kong" flu, whose successor strains predominate to this day and continue to produce, in the United States, more than 20,000 "excess deaths" a year).

A mild and more perplexing pandemic took place in 1977. Dubbed the "Russian flu," this H1N1 strain was identical in all its genes to an epidemic flu strain that circulated in 1950—a medical puzzle, since flu genes are infamously shifty. Some scientists speculate that the virus escaped from a laboratory somewhere in northern China where it had been preserved in a frozen state. The 1977 pandemic did not lead to higher death rates, as most do, nor did it supplant the preceding strain, thus canceling out the scientific dogma that only one flu subtype could circulate at a time. It has been quietly cocirculating in the shadow of H3N2 to this day.

The Barnyard Theory

Where do novel H or N molecules—the matches that light off pandemics—come from? That question has haunted Robert Webster since the 1960s, when he decided that finding the answer would be his life's work. Webster, a tall, baldheaded, robust outdoorsman, grew up on a farm in New Zealand, surrounded by domestic animals and wild landscapes. After studying viral diseases of domestic animals, he moved to Australia in the late 1950s to pursue a Ph.D. There he was assigned to study the influenza virus and its attendant mysteries. He soon discovered that migrating bird populations were rife with asymptomatic influenza. Over the next decades he paid visits to geese in southern China, ducks in Iceland, terns on Delaware's Cape May, muttonbirds in Australia, penguins and gulls in Antarctica, to perform the lowly task of collecting fecal samples for lab analysis. "To really understand influenza in a bird population," he said, "you have to study it all year long, at regular intervals."

More and more, human flu viruses seemed genetically similar to viruses in aquatic birds. By the 1960s, refining scientists' suspicions at the time, Webster arrived at what he calls his "barnyard theory": that the viruses in human pandemics recruit some of their genes from flu viruses in domestic birds. Even his Australian National University colleague, W. Graeme Laver, was at first skeptical. "All my colleagues said, 'You will never, ever be able to show that an animal influenza virus transmits to humans and causes disease,'" Webster told me, heat in his voice. "'You're wasting your time.'" Until the early 1970s, no one had ever searched for flu viruses in wild birds. To test the theory, Webster and Laver organized an expedition to the remote and exquisite coral islands of Australia's Great Barrier Reef, finding antibodies to flu viruses in the blood of nesting sea birds. Several years after the 1968 influenza pandemic, with funding from the National Institutes of Health, Webster, who is now at St. Jude Children's Research Hospital in Memphis, and Laver used the tools of molecular biology to prove that the culpable virus was, in fact, a human–bird reassortant—a genetic hybrid of human and avian viruses, incorporating the hemagglutinin of an influenza virus from a duck in central Europe. In 1972 Webster made his first research expedition to China, which became the epicenter of his work. Over the years, more evidence accumulated.

We now know that the primordial source of all flu strains is migrating aquatic birds. Wild ducks, geese, terns, gulls: these ancient creatures, which have populated the earth for at least 105 million years, harbor the full spectrum of flu viruses (so far as has been identified), H1 through H15, N1 though N9. Domestic avian species—chickens, turkeys, quail, pheasants, geese, ducks—also maintain a large reservoir of flu strains. With a few chilling exceptions, bird flus are harmless to birds, a state of host/pathogen equilibrium that suggests the virus has perfectly adapted to its host over the years, and that even the slightest nucleotide change offers no selective advantage. Ecological stasis also means that, in migrating waterfowl, flu viruses never die out; they simply keep reinfecting birds. All subtypes that have ever arisen or ever will are preserved in perpetuity, like books forever circulating in an open-stack library.

Flu is not primarily a respiratory disease in birds. In ducks, most flu strains replicate in the cells lining the intestines, as well as in the lungs and upper respiratory tract. Passing along the digestive tract, the viruses are then shed in feces. Studies of wild ducks in Canada from

1975 to 1996 showed that up to 20 percent of juvenile birds were infected with flu virus just before taking off for their southern flyways. Through fecal material in lake water, the flu virus apparently passes among feral ducks and is then transmitted to mammals as well as domestic birds, pushing along viral evolution. The passing presence of wild ducks is known to have spawned flu outbreaks in seals, pigs, horses, and domestic turkeys.

Different influenza subtypes infect different animals. Humans are vulnerable to H1, H2, and H3, pigs to H1 and H3, horses to H3 and H7, and so on. This "host range restriction" is determined in part by the virus's hemagglutinin, which recognizes and binds only to specific kinds of molecular structures known as receptors on the surface of cells. Because different species have different kinds of receptors, they can become infected (with few exceptions) only with specific subtypes of virus. The assumption has been that, if we are hit by flu, we pretty much know what menu it came from.

In this vast gene pool, how do new flu viruses make their way to humans? One way is to boldly leap across species. Scientists have collected several examples of this rare event. In 1988, for instance, a healthy 32-year-old Wisconsin woman in her third trimester of pregnancy attended an agricultural fair—and died of an H1N1 virus that had sickened some pigs on display there. In 1991 a Maryland animal caretaker also died after being exposed to sick pigs. Slaughterhouse workers frequently exhibit antibodies to swine flu viruses, even if they don't get sick. But pigs are not the only species to dole out influenza viruses to people. The H7N7 virus in harbor seals, for example, can cause conjunctivitis in humans. Still, when humans get infected with flu viruses from other species, for reasons not yet fully known, the infected person seldom spreads the disease to others.

More worrisome, from a public health standpoint, is the mixing of human and animal flu strains. As Webster showed, one of the hallmarks of its evolutionary brilliance is that the flu virus, with eight segmented strands of RNA, is especially prone to swap genes with another strain of flu when both infect the same host. This viral shuffling-of-the-deck produces scary hands, including the antigenic shift that led to two of this century's major pandemics, in 1957 and 1968. Webster's barnyard theory also points to the most likely locale for this viral mixing and matching: the pig, which, possessing two kinds of receptors in its trachea, is uniquely susceptible to both human and avian flu viruses.

Scientists have actually made a kind of time-lapse recording of this process. In 1979, pigs in Europe became infected with an avian H1N1 virus, a strain quite different from any previous H1 strain that had ever infected humans. By 1984 that virus could latch onto both human and avian receptors, but after 1985, the virus recognized only human receptors. In other words, while multiplying in pigs, the bird virus acquired a preference for human receptors in the pig's respiratory system. Throughout Europe, strains of human H3N2 have also made comfortable homes in pigs. If the replication machinery in the internal genes of these human strains reassorted in the pig with the novel surface proteins of the avian H1N1, it could spell trouble—as in pandemic trouble.

It may well have been in a pig that the deadly 1957 flu (H2N2) swiped two surface genes and an internal gene from a bird virus, while holding on to the remaining five genes from the then circulating human strain. And it may have been in a pig that the 1968 pandemic flu (H3N2) got its hemagglutinin and an internal gene from an avian donor, while retaining the neuraminidase and five other genes from the H2N2 then in circulation.

Any place that is a teeming home to both wild and domestic birds, and to mammals (including humans), may be an ideal breeding ground for a flu pandemic. Europe in the Middle Ages, where humans and animals lived in intimate proximity, was such a milieu; the flu epidemics of that time were probably of local vintage, not imports over the Silk Road. Today's presumptive ground zero for novel flu strains is China. Since 1850, most flu pandemics have originated there. In China all known flu A subtypes reside in ducks and in the water they inhabit or fly over. China also has the world's largest pig population, and one-fifth of the human race, densely crowded even in rural areas (indeed, the Chinese ideograph for "home" is a roof sheltering a pig). China, with its wide climate variations from north to south, hosts human infections year round. And in the great rural expanses of China, pigs and ducks regularly hobnob. Ducks are an important adjunct in farming; swimming around in rice paddies, they nibble on insects and crabs but ignore rice grains, a neat ecological arrangement. And pigs are privileged members of farm households, a gauge of a rural family's wealth. Litters of piglets may spend cold nights inside a farmhouse sleeping alongside family members. On small farms, pigs not only root in the soil for food, inhaling avian flu viruses from wild duck droppings, but are also treated to table scraps that may contain human flu viruses.

Thus, China provides the most opportunities for pigs to be co-infected with human, avian, and swine flu strains—and for flu strains to reassort into unexpected, possibly pandemic, contagions. "Influenza research is a continuing detective story, with all the intrigue of an Agatha Christie novel," writes Kennedy Shortridge. "In the case of a pandemic virus, the H subtype of the prime suspect may be as unlikely a culprit as the vicar's wife." In China, Shortridge adds, influenza is always lurking; it cannot properly be called an "emerging" infection. As he presciently wrote in 1995: "'Elusive' might be more apt."

H5N1

"This is how it begins." When Keiji Fukuda, chief of the epidemiology section of the Centers for Disease Control and Prevention's influenza branch, hung up the phone after talking to his boss, those were the first words to crystallize in his mind. When a severe epidemic breaks out in the United States, Fukuda's job is to find out precisely where the outbreak started, how it spread, and how it can be corralled. If a pandemic were to erupt anywhere in the world, he would head the CDC team dispatched there, lending his and his group's expertise in answering the scientific questions while helping navigate political waters during the emergency.

Fukuda has the right personality for a job that drops him into medical maelstroms: quiet, inward, a good listener, a careful speaker whose words seldom stray outside a narrow temperature range. His close-cropped hair, round wire-rim spectacles, and refined features lend him an almost monkish aspect, a sense of calm in the midst of chaos. His mother was born in Japan, his father in San Francisco to a Japanese family. Himself born in Japan, Fukuda grew up in the picturesque town of Barre, Vermont, where as a child he got fired up reading Kipling's books about India. Three times during college and medical school he decamped for a year to backpack around the world. During one of those jaunts he became interested in malaria, with its astounding human toll, complicated biology, and tricky international politics. Joining the CDC was Fukuda's way of using both his brain and his conscience. Public health crises, he says, "are not intellectual issues that exist in a vacuum."

In the middle of August 1997, Fukuda was on one of his annual sojourns to San Francisco for a few weeks of clinical work, part of his Public Health Service commitment, when he received the call from Helen Regnery, acting chief of the CDC's influenza branch. An H5N1 virus—a pure and devastating avian flu strain—had killed a healthy three-year-old boy in Hong Kong. Regnery didn't need to spell out the implications: the fuse for the next deadly global wave of human influenza, the fourth in the waning century, may have just been lit. He had better pack his bags.

Fukuda, 42-years-old at the time, had imagined this possibility, and in passing reveries of ambition was even eager for it. The last bona fide flu pandemic had occurred almost 30 years before, in 1968, with the worldwide debut of the H3N2 strain. The last scare had been the swine flu debacle in 1976. After Regnery's call, and what ensued over the next half year, Fukuda would never again idly wish for the career-gilding pressure of a pandemic.

The three-year-old boy had become sick on May 10. Doctors prescribed antibiotics for the fulminant infection in his left lung. But the lung collapsed, the boy's liver stopped working, his brain swelled (from Reye's syndrome, caused by aspirin use), and finally his kidneys gave out. On May 21, he died. The boy's virus was some kind of influenza A virus—but exactly which kind, the Hong Kong Department of Health lab couldn't determine. Now, three months later, a group of researchers in the Netherlands identified the viral culprit as an unre-constructed bird flu; the CDC confirmed the finding.

If true, the lab reports were both puzzling and alarming. The think-ing had been that in order for a person to become infected with an avian flu strain, the viruses must first be passed through a pig. How did a bird virus jump directly to a human—and kill? Had there been other undiagnosed cases? Was something new going on in the tangled ecology of flu?

International media jumped on the story. The boy's death was a "significant event," Robert Webster told one news-service reporter. The World Health Organization's surveillance effort for both humans and animals in China was stepped up. Researchers began working on a rapid diagnostic test for the new strain. Among flu experts around the world, the betting was that this was a case of laboratory contamina-tion, nothing more. Fukuda quickly ruled out that comforting scenario. He also learned that the boy had always been healthy, and that no evidence showed the virus spreading through the population—no rise in deaths had occurred, no increase in the number of patients on respi-rators, no surge of feverish cases in emergency rooms, no jumps in other suspicious diagnoses.

To find out whether H5N1 infections had been lurking unnoticed, the CDC team tracked down blood already collected for other medical reasons, including blood from healthy donors and from children par-ticipating in studies unrelated to flu. To see whether the boy had passed on the virus to his family and others, or whether he had contracted it from someone else, investigators drew blood from everyone who had had contact with him. Finally, the team also drew blood from poultry workers. Earlier that spring, before the boy died, H5N1 infections had broken out at three chicken farms in the rural northwestern part of the New Territories (the land north of Kowloon). On two of these farms, the mortality rate among the birds was 100 percent. Authorities or-dered the rest of the birds slaughtered. Exposure to poultry offered a tantalizing possibility of how the boy might have become sick. In his classroom was a "feathered pet corner," where chicks and ducklings had died in the days just before the boy became ill.

By the end of the investigation, about 950 samples of blood had been drawn. Only seven contained antibodies indicating exposure to H5N1—five (of 29) from poultry workers, one from a classmate of the dead boy, another from a doctor caring for the boy. Of these, only the doctor recalled feeling under the weather that spring—but it was too late to prove he had actually been sickened by H5N1, much less that he had caught the virus from his young patient. Scrapings from the pet corner were negative.

By early September, after Fukuda had been in Hong Kong three-and-a-half weeks, nothing seemed clear. The scientific consensus was that the boy's death was aberrant and inexplicable, but that it didn't pose a problem to others because it hadn't reassorted with a human strain of flu. Still, Fukuda kept pointing out to blasé colleagues that although H5N1 didn't seem to be an imminent pandemic threat, it *was* a novel virus in humans. He sat back on the plane home, exhausted and uneasy. He had never found out how the boy had become sick. But he had seen, up close, how quickly rumors can circulate, headlines can plant panic, and institutions can be strained, even during a false alarm. If H5N1 had turned out to be the start of a new pandemic, the world wasn't ready for it.

***

On November 6, 1997—six months after the three-year-old boy had died and two months after Fukuda had flown back to his office in Atlanta—a two-year-old boy in Hong Kong came down with a fever and sore throat and cough. The next day he entered the hospital, and doctors swabbed his nose for a sample. The boy recovered and was discharged two days later, but when lab workers tested the sample, it turned out to be H5N1. Case #2. The avian flu had once again jumped species, without a pig intermediary. Where had it hidden during the six-month lull? No one knows. Perhaps the virus had thinned out when the poultry infected in rural Hong Kong were slaughtered that spring. No infected chickens meant no chance for the virus to jump to humans. But even if domestic poultry had been temporarily eliminated as a reservoir, the virus must have continued to thrive somewhere.

The CDC heard about the case in late November. On the night of December 6, Fukuda and a colleague in the CDC's Epidemic Intelligence Service (EIS) arrived in Hong Kong. A medical officer from the Hong Kong Department of Health greeted them at the airport with more bad news. Just that day, H5N1 had turned up in a specimen from a 13-year-old girl who was on a respirator. Worse, a 54-year-old man had died of respiratory failure, and H5N1 was isolated in him as well. He had been sick since late November, when he first felt fever and chills, but had gone on a vacation bus tour in South Korea. On November 29 he was admitted to the hospital with severe pneumonia. The doctors were helpless. Case #4. "The second investigation started out at 100 miles an hour," Fukuda says. "And it got faster and faster."

It looked like the epidemic everyone had braced for in the summer. On December 4, a 24-year-old Filipina domestic helper suddenly felt feverish, headachy, and dizzy; five days later she was in critical condition with respiratory and renal failure. On December 7, a feverish five-year-old girl—whose grandmother was known in the neighborhood for scavenging dead chickens from the markets—began vomiting; three days later, she was admitted to the hospital. In a bird-worshipping culture, the new avian flu threw the city into a panic. Residents dumped pet cockatoos, pigeons, and mynah birds in front of the Society for the Prevention of Cruelty to Animals. Hong Kong's biggest poultry market was closed for sterilization. Robert Webster speculated before the press that if a pandemic were really afoot, "There is nothing we can do to contain it." Even the city's director of health sounded a note of defeat. "It seems we are entering a competition with the virus," she said. "We are working at breakneck speed."

On December 12 events took an ominous turn. A sick two-year-old cousin of the five-year-old girl tested positive for H5N1. Shortly after, so did a four-year-old cousin. Investigators became obsessed with one question: Was the virus jumping from person to person? Word came in of another chicken farm outbreak—the fourth—in the New Territories. With each new development, the media bore down harder and harder, and the public grew more alarmed. Why were some people getting sick and not others? Scientists already suspected that the virus was transmitted from poultry to humans. But as new cases kept surfacing, Fukuda wondered: Are other animals harboring the virus and spreading it to people? Mice? Cats? Nothing seemed too farfetched. To get answers, investigators had to interview the patients' families and other close contacts. They often found that the aggressive Hong Kong media was one step ahead of them—besieging the relatives of infected patients with microphones and cameras. By the time EIS officers arrived to collect crucial details about victims' activities and exposures, families refused to speak.

On December 21, Case #3, the 13-year-old girl, died. On December 23, Case #12, a 60-year-old woman, died. The tally was beginning to suggest an epidemic-in-progress. Off the record, health officials told reporters the virus was being transmitted between people. On Christmas Day, Fukuda's mood turned black. Hospitals were reporting more flu patients. Fukuda's mind kept returning to what he had read about the Spanish flu—the suddenly accelerating deaths that autumn, marking what in retrospect could be seen as the second wave. The pattern seemed "eerily reminiscent of 1918," Fukuda said. "I thought this thing was beginning to explode."

On the eve of the regular flu season, which begins in February in Hong Kong, public health officials also feared that H5N1 would swap some of its genes with the well-adapted H3N2 or H1N1 human flu viruses that were circulating, or with human viruses infecting pigs, creating a novel killer that would easily spread from person to person. Reassortment, of course, had ignited the 1957 and 1968 flu pandemics. It was a race against viral evolution. What scientists knew about the past behavior of lethal avian flus compounded their fears in Hong Kong. Influenza rarely attacks domesticated chickens, because they are sheltered indoors, away from the strains spread by feral ducks and other birds. Two previous epidemics in chicken farms, however, had showed how savage an avian flu can become away from the wild. The first took place in April 1983, when an H5N2 virus spread in vast commercial chicken operations in Pennsylvania. At first, the disease was mild. By October, it had become vicious. The chickens had swollen, bluish combs. Their eyes and legs were bloody. They laid eggs without shells. Autopsies showed that the virus produced hemorrhaging in every type of tissue, including the brain. In 1993, H5N2 broke out in Mexican chicken farms. It too started with deceptively mild symptoms and ended with decimation.

The emerging Hong Kong H5N1 virus had stretches in its hemagglutinin protein identical to those in the Pennsylvania avian flu. And as University of Wisconsin virologist Yoshihiro Kawaoka discovered, the virus's lethality was not limited to birds. When he injected human H5N1 isolates into mice, the virus replicated in all organs, virtually dissolved the lungs, and killed the animals. "This is the most pathogenic virus that we know of," he says. "One infectious particle—one single infectious virion—kills mice. Amazing virus." Could it do the same to humans? Public health officials worried that it might just be a matter of time before the virus adapted to people, mutating or reassorting in a way that reduced its human hosts to a bloody pulp.

On Sunday, December 28, the Chinese government announced that it would kill all the birds on chicken farms, wholesale markets, and live bird stalls in the territory—more than 1.5 million creatures over a three-day period—and would temporarily halt imports of birds from China. Only six months before, Hong Kong had been released from British colonial rule, and for weeks politicians had been blaming the leadership for dragging its feet. The dramatic gesture was partly meant to build confidence in the new regime, but it also sprang from the science. Just before the holiday, Robert Webster and Kennedy Shortridge had organized a team to begin testing birds in the city's "wet" markets, where sellers washed down stalls and sidewalks twice a day with water. They found that 10 percent of chickens in the markets harbored H5N1. The virus had heavily infiltrated wholesale markets and farms. Ducks and geese also carried the virus, though these species showed no symptoms of disease.

On December 29 the slaughter began. Scuffles broke out between chicken vendors and reporters, whom the vendors blamed for exaggerating the problem. Buddhist monks held a marathon seven-day prayer chant for the souls of the dead birds. Workers from the Agriculture and Fisheries Department—many of them deskbound functionaries—donned white coveralls and surgical masks to carry out their assignment, loading the birds into black plastic bags and pumping in carbon dioxide. When the CO_2 ran out, they drew knives across the birds' throats. With feathers and dust flying, reporters crowded in close wielding cameras and notepads. Members of the EIS team, in a smug reverie, designed a study in which they would collect timed blood samples from the intrepid media, tracking the rise in antibody levels.

Though Keiji Fukuda couldn't know it at the time, the December surge in flu cases merely marked the onset of Hong Kong's annual influenza season: routine cases of previously circulating strains. The last person to be diagnosed with the strange bird flu, H5N1, became ill on December 28, the day before the slaughter commenced. This 18th hospitalized victim was a 34-year-old woman with several underlying conditions. She became ill with the flu, then pneumonia, then a mean, antibiotic-resistant form of blood poisoning. In early January she died. With her death the H5N1 outbreak finally ended. But in Hong Kong, the story was not quite over.

Time Travelers

Like the 1918 flu, the brief H5N1 epidemic primarily struck down healthy people. Even modern medicine's most powerful artillery couldn't stop the virus. This helplessness revived a question that had long preoccupied researchers: What makes some flu viruses so catastrophic? Long before H5N1 broke out, many researchers had become convinced that if they could find a specimen of the 1918 flu virus, they could take it apart and discover what made it—and what could make future pandemics—so deadly.

Jeffery Taubenberger is not a flu man. Chief of the division of molecular pathology at the Armed Forces Institute of Pathology (AFIP), in Washington, D.C., a down-at-the-heels government lab, his primary interest is immunology, specifically how stem cells—the precursors to all kinds of blood cells—develop in bone marrow. "I'm interested in projects that go at big questions," he said, "knowing that the only way to get at big questions is to do one tiny thing at a time."

In science, an inspirational spark in one specialty can make a spectacular leap to a seemingly distant domain. For Taubenberger—a short, sloe-eyed, eloquent man who composes chamber music in his off hours—the creative path to the 1918 flu virus began in 1987 on the New Jersey coast, with a die-off of bottlenose dolphins. The epidemic spread south to Florida, finally petering out in 1988. At least 10,000 animals died. At first, experts blamed the epidemic on red tide, an overgrowth of toxic algae. But when veterinary pathologists looked at the dolphins' tissues, they concluded that the damage bore the hallmarks of a morbillivirus—a marine relative of the virus that causes distemper in dogs, and the virus that causes measles in humans. After 1987, other sea creatures—seals, porpoises, various species of dolphin—suffered similar die-offs around the world, making morbillivirus a true emerging infection in the aquatic realm. Taubenberger and a colleague wanted to track the virus's spread in the Atlantic and the Pacific. To begin that detective work, a colleague in Taubenberger's lab spent more than a year refining a technique for extracting the virus's fragile RNA from tissues that were sea-logged and badly decayed.

The breakthrough solidified the lab's unusual, if unheralded, expertise—getting genetic material out of decidedly unpristine tissue. These creative fires were stoked by a practical necessity. Unlike hospital laboratories, which analyze fresh samples, the AFIP traffics in old specimens. The institute's predecessor—the Army Medical Museum—had been launched by order of Abraham Lincoln during the Civil War, with the far-sighted goal of collecting specimens that could be used in research on military medicine and surgery. After all, in the War Between the States, as in virtually all wars before and since, more combatants died from infectious disease than from battle wounds.

Today, the National Tissue Repository, a division of the AFIP, houses three million specimens from autopsies on soldiers and clinical records describing the progression of their diseases. Some of these samples are organs sloshing in formalin solution and packed in thick see-through plastic bags. Some are translucent slices pressed between glass slides. And some of the oldest are slivers of tissue about the size of a thumbnail, preserved and then embedded in small blocks of par-affin wax—slivers that can still be sliced thin and peered at under a microscope. During the early 1990s, the molecular pathology lab at AFIP had quietly cornered a niche market, perfecting molecular tech-niques that could retrospectively diagnose infectious and genetic dis-ease in these old specimens.

Once Taubenberger had proved it's possible to retrieve small amounts of DNA and RNA from fixed tissue, he searched for a project that would showcase his lab's exotic expertise. Perhaps they could ge-netically analyze tissue from victims of what in the Civil War was called "camp fever." Was it caused by malnutrition? Or by typhoid, a bacte-rial infection caused by a strain of salmonella? When someone men-tioned the 1918 flu, Taubenberger had an "Aha" moment. The AFIP has perhaps the world's largest collection of tissue samples from vic-tims of the pandemic: about 120 specimens from American soldiers who died of the disease. If Taubenberger's lab could genetically se-quence the virus behind that global catastrophe, it would go down in history not only as a technical feat but could presumably lead to un-precedented insights into the biology of flu.

The first step was to find tissue samples that harbored the actual virus, not just vestiges of bacterial pneumonias caused by staph, strep, and haemophilus bacilli. It wouldn't be easy. The flu virus is a hit-and-run germ, rapidly copying itself in host cells for only a day or two after the infection, then leaving the scene of the crime. By the fifth day after infection, the virus has disappeared from lung tissue. Taubenberger was therefore searching for victims who died less than a week after first feeling ill. In an underequipped office, he jury-rigged a micro-scope to view microfiche copies of the World War I soldiers' clinical records. Peering at the pale, spidery script of his grandparents' era, he tried to discern which soldiers had succumbed to massive pulmonary edema and hemorrhage, the classic signs of acute influenza death.

Taubenberger selected the most promising samples and gave them to Ann Reid, a gifted lab technician. Reid was a true intellectual peer—every bit as methodical and creative as Taubenberger, and possessed of what bench scientists call "good hands": the rare ability to handle minute quantities of materials swiftly and adeptly and with fanatical precision, setting up the same experiment again and again and again. But in the rigid class hierarchies of science, Reid—a striking woman with a direct gaze and an easy manner—had little standing. Not only did she lack a Ph.D., but she was a part-time employee, arriving at 6:30 each morning and leaving at 2:30 in the afternoon. "I *have* to leave," she says. "Because I have kids to pick up from school and soccer practice and piano lessons, and I do a lot of church work. There's a whole life that takes over when I walk out this door."

Reid's job was to employ a technique known as RT-PCR, for reverse-transcriptase polymerase chain reaction. PCR lets scientists make millions of copies of known stretches of DNA, the ladder-like molecule that guides all biological functions. A variation of this technique, RT-PCR enables researchers to amplify RNA viruses such as the influenza virus by adding a step that converts their RNA to a form of DNA.

PCR and genetic sequencing is an endlessly recursive process. In a nutshell, Reid's work went something like this: She started out with a tiny quantity of the virus's genetic material from the AFIP's most promising specimens. Then, using the benchtop PCR machine, she tried to amplify a particular nucleotide sequence that barely changes across flu strains. If she found this conserved stretch of genome, she would once again amplify it, spell out its nucleotide sequence, and confirm the finding.

For 15 months, from March 1995 to July 1996, Reid labored exclusively on the 1918 flu project and did not find a single shard of the virus's genetic material. Then she sat down with a new sample. This time a positive result came back. Reid and Taubenberger went to a computer and entered the 50-odd base pairs into GenBank, which compared it against all other published sequences. The computer digested their request for an endless minute and flashed this response: The sequence belonged to an influenza matrix gene which, though similar to strains from the 1930s, did not perfectly match them. This meant the virus was unique. It was old. It could only have come from the 1918 flu.

Roscoe Vaughan, a 21-year-old army private at Fort Jackson, South Carolina, took ill on September 18, 1918, just after the start of the deadly fall wave. On September 26 he died. An autopsy revealed that, while his left lung showed typical signs of massive bacterial pneumonia, his right lung was inflamed only at the ends of the bronchioles and their surrounding air sacs. It was a snapshot of the influenza virus vigorously replicating in the lungs. Apparently, the right lung lagged several days behind the left in the disease process, and that unusual asynchrony worked in Taubenberger's and Reid's favor; of the first batch of samples examined under the microscope, only Vaughan's contained the flu virus.

Reid started figuring out the virus's gene sequence. So badly deteriorated was the virus's RNA, she had to work with segments usually no more than 130 base pairs in length—in a virus that contained more than 15,000 base pairs among eight different gene segments. Reid designed primers—chemical frames—that would bind to overlapping genetic fragments, allowing her to painstakingly assemble a picture puzzle of the genome.

When Taubenberger and Reid found identifying snippets of five different genes, the highly variable hemagglutinin and neuraminidase genes tipping them off to the lineage of the virus, they decided it was time to publish. In September 1996 they sent their first paper to *Nature*. The manuscript came back instantly—rejected, unread. They sent it to *Science*. Same story. They sent the article to a third publication whose editor was interested—so interested that when he lunched with an editor at *Science*, he mentioned the manuscript. The *Science* editor's ears perked up. "I think that came across my desk," she said. She called Taubenberger and asked him to resubmit the paper. Six months later, in March 1997, the article in *Science* described the first cut at genetically sequencing the 1918 flu virus and setting it in an evolutionary scheme. The authors concluded that the virus was indeed an H1N1 virus, similar to that in swine flus.

For the first time in Taubenberger's career, his work got noticed. It was a surreal time for the 36-year-old. Twenty phone calls a day from reporters. TV interviews. Documentary films. Tensions flared in the lab, disgruntled colleagues feeling Taubenberger was getting more than his fair share of accolades. Flu experts from outside regarded Taubenberger as either interloper or parvenu.

In the glare of publicity, Taubenberger couldn't ignore a problem on the horizon. The archival specimen was tiny—about the size of a thumbnail. To fully sequence the 1918 flu virus, he would need a fuller sample. Five months after their paper was published, Taubenberger and Reid recovered a second positive paraffin-embedded sample from the AFIP collection. While this second case was a boon, it was not enough. With the finish line in sight, Taubenberger and Reid would run out of material before they could reach their goal.

***

In the spring of 1997, living in San Francisco, Johan Hultin came across an article that described Taubenberger's dramatic breakthrough in decoding the 1918 flu virus. It seemed like the culmination of his own dream—a blueprint for the virus, if not the virus itself. Not until two years after his midcentury Brevig Mission trip was the genetic code elucidated. Not until decades later could his colleagues have known that the RNA molecule making up the influenza virus is too fragile to withstand Arctic temperatures. Hultin had abandoned his dreams of becoming a great virologist and settled for a medical degree, a career in pathology, and raising a family. Along the way, he was constantly plunging into projects that tested the limits of his physical strength and literally salvaged the hidden remnants of history—locating a wagon abandoned by the ill-fated Donner Party, climbing an uncharted peak in Pakistan, restoring an ancient coastal labyrinth in Sweden.

In July 1997, he wrote to Taubenberger: "Your article in *Science*, March 21, 1997, and your discussion with Jim Lehrer I found most interesting. . . ." As "an entirely private undertaking," Hultin offered to deliver to Taubenberger specimens from the frozen victims at Brevig Mission. He had once gained permission to enter the mass grave, and he was confident he could do it again.

Taubenberger had taken time off after the birth of his second child, and hadn't seen his mail for several weeks. When he read the letter, he immediately phoned Hultin.

"When can you go?" Taubenberger asked, expecting that Hultin's trip would take a year or so of planning.

"I can't go this week. But I can go next week," Hultin responded, feeling sheepish about the delay.

Taubenberger said nothing for a moment. He couldn't believe his luck—that out of the blue, a stranger with Hultin's credentials was offering his services. Hultin thought the silence was skepticism. He was grateful Taubenberger didn't hang up.

August 19, 1997, was windy and rainy in Brevig Mission. In a community that was 92 percent Inuit, the majority unemployed, Hultin was a sight: a six-foot-tall Swede with white hair swept back and a white goatee—a kind of lean Colonel Sanders, without the treacle. At 4 p.m., dressed in waders and suspenders and his wife's white floppy sun hat, he began removing squares of sod. Four young village men in jeans and baseball caps helped out. By 10 p.m., the light of the long Arctic summer still strong, they still hadn't reached permafrost.

The next day, their shovels hit an implacable layer of otherwise dark, normal-looking ground, at a depth of four and a half feet. Permafrost at last. It was a good sign, suggesting that even in high summer, the earth had thawed only near the surface. The yield was less than spectacular—only the remains of a child and an adult, too decayed for even small tissue samples.

On August 21, at a depth of seven feet, shovels began to expose a body that looked far better than the others. The skin over the chest was hard and tough, an inch thick. When Hultin cut into it, he saw a yellowish substance with whitish gray specks—decayed fat. It was a woman between 25 and 35. A black tuft of hair sprouted from her skull. She was turned slightly to the right in her earthly resting place, an almost demure pose. Most important, from Hultin's point of view, her lungs were not only intact but also deep dark red in cross-section, suggestive of the hemorrhaging in acute influenza death.

The next morning, Hultin used his pruning shears to open the rib cage and remove the dry, leathery lungs. On a plywood board that had washed up on the beach, he cut slices of the lungs, placing some in formalin (for later work on cell pathology), others in either alcohol or a solution of gaunidinium thiocyanate, which saves genetic material while inactivating any live organisms—both fixatives that would preserve the virus's precious RNA.

Later that day, resting on an overturned pail, Hultin was absent-mindedly gazing at the open grave. All day long he had been turning over the same question in his mind: why was this woman's body so beautifully preserved, unlike virtually all the other corpses, even the bodies closely flanking her, deep in the permafrost. Of 11 bodies found in the grave, only four had retained soft tissue—but none as unspoiled as this. Suddenly it dawned on him: She was obese. A layer of thick fatty skin had insulated her ravaged lungs. He named her Lucy, a tip of the hat to anthropologist Donald Johanson's 1974 find, an ancient australopithecine and possible ancestor of humans. Hultin hoped that "Lucy," a name derived from the Latin word for "light," might illuminate the deadliest infection in history.

After flying back to San Francisco, Hultin meticulously wrapped the remnants of the 1918 flu. He dispatched four packages via separate carriers to Jeffery Taubenberger, ready to wait a month or so for the results. Ten days later he got a phone call from Taubenberger. Lucy tested positive for the 1918 virus. And because this frozen cache was the most abundant sample, it became the new index case against which all other samples would be compared.

***

So why *was* the 1918 flu so uniquely deadly? When Taubenberger began his quest, he hoped that a single genetic mutation would solve that puzzle. Perhaps the 1918 flu had a mechanism akin to the "fowl plague" flu viruses (including H5N1), which permits the hemagglutinin molecule to be cleaved—and thus made infectious—in every cell of the body, leaving a bloody pulp. But Taubenberger found that the 1918 flu virus does *not* have this telltale genetic glitch, or anything else in its surface protein-coding genes that would render the virus pantropic, i.e., capable of attacking many tissues. Far from being a blueprint for the biological equivalent of St. Peter's cathedral, it looks like the plan for a concrete parking garage.

Even when Taubenberger and Reid are able to publish the entire 1918 sequence, by 2004 or so, it may tell scientists nothing. "You can think of sequence like you think of the notes of a score," Taubenberger said. "Music only exists as sound, so that the score to Beethoven's Fifth Symphony is not Beethoven's Fifth Symphony. Only when an orchestra plays it does it become Beethoven's Fifth Symphony. That's just like the sequence of a virus. It's just the directions for making the virus, it's not the virus."

Today it appears that the virulence of the 1918 flu may stem from many factors: its ability to copy itself quickly, to inflict damage deep in the lungs, and to spread with astounding ease. These may result from interactions between the virus's genes or between the genes and the host's immune system. And despite the effort Taubenberger and Reid spent cracking the virus's code, this knowledge may not protect us against the effects of the next pandemic.

All influenza viruses are not created equal. What may explain the differences is a technique just on the horizon: reverse genetics, in which scientists, knowing the full sequence of virulent and nonvirulent viruses, introduce subtle mutations into cloned genes in order to see how these affect virulence and transmissibility in mice or ferrets. Such visionary methods are risky. As Robert Webster put it, "We can't generate 1918-style virus. We have to be careful not to create monsters."

Taubenberger and Reid's subsequent work has revealed a provocative fact: the 1918 flu pandemic was not set off by the same circumstances as the ones of 1957 and 1968. Those viruses had surface proteins that jumped directly from birds, coupled with human-adapted core genes. By contrast, in the 1918 virus, the surface genes are mammalian in character. Though probably originally derived from a bird, the virus had spent years adapting to life in mammals, either pigs or humans. Taubenberger estimates that the virus began making its rounds in humans in 1917, just before the pandemic's spring wave. If that kind of leisurely genetic adaptation happened once, it could happen again. Maybe people pick up animal flu viruses all the time. Maybe pigs aren't the only mixing vessels. Maybe that's the lesson of Hong Kong.

The good news: if people are catching animal viruses all the time, the circumstances behind a pandemic must truly be rare. The bad news: if people are catching animal viruses all the time, and these viruses are capable of hanging around and adapting, a pandemic is perpetually around the corner.

The Next Pandemic

So convinced are public health officials of the inevitability of a flu pandemic that they call the past 30 years' respite the "inter-pandemic" period. A gloomy locution, like calling good health the "inter-disease" period, or marriage the "inter-divorce" period, the phrase nevertheless captures the essentially ominous quality of flu. As Kennedy Shortridge wrote in *The Lancet*: "Put simply, each year brings us closer to the next pandemic."

This certainty has led to national and international influenza pandemic preparedness plans. Though spun in flat bureaucratic prose (the U.S. draft plan calls nasty vaccine side-effects "untoward and unexpected programmatic events"), these documents try to anticipate all the ways a pandemic could unroll and how modern medicine might outflank this microscopic enemy. Pandemic preparedness is provisional in nature, always subject to revision as more quirky virus behaviors come to light.

The year 1918 offers one frame of reference for a modern pandemic. That year, October was a scene out of Dante, in which deaths mounted at an alarming rate. In one week Philadelphia lost 4,500 citizens; the dead were left in gutters. At Fort Gordon, in Atlanta, soldiers' coffins were stacked like cordwood. In Chicago police closed theaters, skating rinks, movie houses, night schools, and lodge halls. In San Francisco the order to don gauze masks outside the home was written into law.

Is that how a 1918-force epidemic would hit us today? It's often said that we live in a world vastly different from that of 1918, that modern medicine—which has cured cancer, scoured plugged-up arteries, and planted fetal cells in aging brains—could surely handle something as trivial as flu. It's true that today's antimicrobials might limit secondary pneumonias, and it's true that doctors can now treat such complications as heart and lung failure. Still, CDC officials estimate that up to 200 million Americans could sicken; 45 million could need outpatient medical care; 800,000 could be hospitalized; and up to 300,000 could die.

Our inability to help those sickened by the recent H5N1 virus—patients who had received excellent medical care—has revealed how shaky our position really is. The truth is, scientists still don't know what biological gears drove the 1918 pandemic. Nor do they know why many who recovered had aftereffects such as Parkinson's disease. Nor can they be assured that in our more densely populated, technology-dependent world, a severe flu pandemic wouldn't be worse than ever. Contemporary air travel would surely hasten the global spread of a new virus, from four months in 1918 to perhaps four weeks today. (Not long after the H5N1 outbreak, Joshua Lederberg wondered how many travelers had passed through Hong Kong's Kai Tak Airport in November and December 1997, when that epidemic hit full force. The answer: more than 4 million.) Urbanization would speed the transmission of the virus once it arrived in a new locale. Today's antibiotic-resistant strains of strep, staph, and other agents of secondary infections would leave us perhaps less prepared than we were in 1957, before antibiotic use was so promiscuous. Infection with the AIDS virus would leave many flu victims with faltering immunity, raising the death rate. And the larger elderly population in the United States would raise the death rate dramatically, a trend borne out in interpandemic years. Since 1985, hospitalizations for all diseases have dropped 32 percent—the legacy of health mainenance organizations. But hospitalizations for pneumonia have risen 50 percent—the legacy of longer life expectancy.

***

All of which underscores the need for vigilant surveillance of novel flu strains. In the World Health Organization pandemic preparedness plan, national influenza centers in more than 80 countries—with their ability to quickly identify unfamiliar viruses—form the front line. If they find a strange new isolate, they send a sample to one of the four WHO Collaborating Centers for Reference and Research on influenza (including the CDC in Atlanta), where the variant can be confirmed and further studied antigenically and genetically.

Still, the politics of close surveillance is tricky. It took Robert Webster decades to build trust with his Chinese hosts, who understandably feel threatened when the West labels them the influenza epicenter of the world. Not illogically, the Chinese have adopted a certain fatalism about influenza pandemics; they know they would be the first to get slammed, and that emergency measures would arrive too late to help. The Chinese also place more priority on other diseases: hepatitis, Japanese encephalitis, AIDS, polio, and rabies. And Chinese scientists resent the fact that their country is a mecca for ambitious researchers from elsewhere in the world, who may fly to Asia solely to snare interesting viruses and earn themselves academic acclaim. Animal flus are perhaps even more difficult to report than human strains, because these diseases taint trade. Despite these problems, the monitoring of new influenza viruses and of human and animal disease is tighter in Hong Kong than anywhere else in the world. Mainland China also has excellent virologic surveillance.

Yet even in the state of red alert after H5N1, things don't always work the way they should. In March 1999 in Hong Kong, for example, it was déjà vu all over again. Two little girls, ages one and four, caught the flu. At first, nothing about their cases seemed to stand out. They had all the usual symptoms—a sudden fever, sore throat, headache, vomiting. And even though both had medical histories that made doctors extra cautious—the one-year-old was very small and had been diagnosed with "failure to thrive," the four-year-old suffered asthma and eczema—both recovered completely and were soon discharged from the hospital. But when WHO laboratories in London and at the CDC analyzed the virus, they discovered that the girls had been infected with the strain H9N2. The pathogen was known to cause mild illnesses in poultry, as well as in pigeons and quail, and was antigenically similar to swine viruses isolated in Hong Kong in 1998. Never before, however, had H9N2 been isolated in people.

When Hong Kong officials announced the findings in April 1999, flu experts around the world were surprised and dismayed. Was this a reprise of H5N1? They were even more startled to hear that in August 1998—a full seven months before the two Hong Kong girls had become sick with H9N2—five China mainlanders had also fallen ill with the virus. Not until after the Hong Kong findings were broadcast did Chinese health officials publicly acknowledge the earlier cases. Initially they may have been reluctant to discuss the findings because the original samples may have been contaminated with avian viruses, a result of shoddy lab technique. News reports quoted the director of the Influenza Surveillance Center in Beijing as saying that he neglected to report the cases to WHO because there had been no signs of a large outbreak. Like the Hong Kong patients, the mainland patients were mostly small children, and all quickly recovered. But so far, no WHO labs have seen the five isolates. That worries Nancy Cox, chief of the CDC's influenza branch. The success of the WHO collaborating laboratory system, she says, depends on "how willing people are to share their information and share their viruses at critical points in time."

That two quizzical bird viruses—H5N1 and H9N2—jumped to humans in rapid succession has unsettled many flu specialists. Although scientific dogma states that humans don't have the right receptors for bird flu viruses, that dogma is starting to crumble. "Common sense tells me that both the H5N1's genome, and the H9N2, have something very special," Webster says. "They have special characteristics that allow them to spread to humans. We think it's related to their internal genes rather than their receptor specificity." Indeed, Webster found that the six internal genes of the two viruses—everything except the hemagglutinin and neuraminidase genes—were virtually identical.

Today, 10 percent of the birds found in Hong Kong's live markets harbor H9N2. And in echoes of 1997, a new and deadly H5N1 strain in chickens forced the Hong Kong government in 2001 to order the slaughter of 1.2 million market birds. In Webster's mind, that fact makes intensive surveillance even more critical. In human blood, what's the baseline of antibodies to animal flu viruses? Are antibodies to, say, H9N2 rising from baseline? If so, are the viruses somehow changing and adapting to humans? Will H5N1 stage a comeback? "If we could get a handle on that before it really goes," says Webster, "then we have an advantage in making a vaccine that would head off a pandemic."

But Ed Kilbourne, the grand old man in American flu research, is skeptical that we can ever track a nascent pandemic. No matter how influenza smolders theoretically, we won't know it's caught fire until people get sick. Perhaps lethal genetic reassortment is a rare and idiosyncratic and ultimately unpredictable event. "To me," says Kilbourne, "what's happened with H5 and H9 strengthens the case that you can't just have 'alarums and excursions.'"

Kilbourne is the battle-scarred veteran of one such false threat—the February 1976 swine flu scare at Fort Dix, New Jersey. Private David Lewis, a 19-year-old recruit, previously fit and healthy, went on a forced five-mile night march while suffering what seemed like a bad cold. On February 4, he collapsed and died of pneumonia. As it turned out, the pneumonia was the aftermath of an H1N1 infection—a strain found in domestic swine, and the same subtype as the 1918 flu. Lewis was one of five military recruits from whom H1N1 was isolated. Ultimately, hundreds of recruits turned up antibodies to the virus, though only 12 became sick. The same day the Fort Dix cases were identified at the CDC, the *New York Times* op-ed page ran an urgent entreaty by Kilbourne. Declaring the dogma of the time about 11-year pandemic cycles, he predicted the next onslaught around 1979: "[T]hose concerned with public health had best plan without further delay for an imminent natural disaster."

That February, the disaster seemed to have arrived early. Anticipating a new pandemic by the next winter flu season, U.S. officials ordered emergency vaccine production under government contract and a mass vaccination campaign. "Fiasco" is the word invariably appended to what happened next. First, the issue of liability stalled vaccine makers. Then, one manufacturer used the wrong virus in formulating the vaccine. That summer, a deadly outbreak at a convention of American Legion members raised panic that the new swine flu had finally appeared (the outbreak turned out to be the first known appearance of Legionnaires' disease). Delayed by bureaucratic hurdles, the vaccination program started late. Eventually, 45 million Americans received the drug—only to learn that the inoculation raised their risk of contracting Guillain-Barré syndrome, a neurological disorder. In the middle of December, beset by controversy from the start, the vaccination program was finally terminated.

Meanwhile, no swine flu epidemic had erupted. Twenty years later, Walter Dowdle, who had served as the CDC's laboratory chief during the swine flu scare, wrote: "It has been said that the pandemics of 1957 and 1968 were diseases in search of a vaccine. In 1976, the vaccine was in search of a disease."

For public health officials, the lesson from Fort Dix has been that before you declare a novel virus pandemic, make sure the microbe has legs and can spread efficiently from person to person. A pandemic cannot truthfully be deemed "imminent" unless a new flu virus has caused unusually high rates of illness and death in several geographic areas, whether in the same country, several countries, or on more than one continent. The "swine flu affair" also brought up another policy dilemma: the more science uncovers potential risks, the more public health officials and government administrators must ponder how to politically address those risks—risks magnified by scare-hungry media.

***

Once a pandemic has been unleashed, the best way to curb its effects will be a vaccine. Yet formulating an effective vaccine in a hurry is akin to building a World Series contender on short notice: all the parts must work together superbly—and even then you need some luck. Two-and-a-half years after the H5N1 "index case"—the three-year-old boy—died, the U.S. Food and Drug Administration still hadn't approved a vaccine against the strain. The virus, which had already decimated chicken farms, also killed the embryos in laboratory eggs.

In a pandemic, pharmaceutical companies would have to embark on a crash course in producing a new vaccine, the technical difficulties of which can't be known in advance. For more than 50 years, flu vaccine has been made by inoculating seed virus in embryonated hen's eggs, purifying and chemically treating the harvest, knocking out the infective elements, and adjusting the concentration against reference biological standards. In some cases, because new flu strains don't grow well in eggs, so-called high growth reassortant viruses are used. These combine the hemagglutinin and neuraminidase genes of the new strain with remaining genes from strains known to reproduce quickly in eggs. Huge vaccine-ready flocks—five million birds—must be hatched six months in advance, so that they're mature enough to lay eggs. It takes at least half a year to prepare the annual flu vaccine, which is used only in a small proportion of people, such as the elderly or immunocompromised, who are at higher risk. In a pandemic, public health officials may need to turn to manufacturing systems that don't depend on eggs—cultivating the suspect virus in tissue or using recombinant DNA techniques, which replace lethal stretches of the genome with benign parts.

The tenuousness of vaccine manufacture became clear in 2000, when one of the new strains making up the annual flu vaccine—a strain known as Panama A—failed to grow robustly in eggs. With this unexpected glitch, the United States caught a brief glimpse of pandemic terror. The shortfall became public that June, at a time when health officials had expected that record numbers of Americans would seek immunizations; the CDC had just lowered to 50 the age at which all individuals were advised to get a flu shot. Sure enough, flu hysteria took hold. A "gray" market in the vaccine took off, with secondary wholesalers charging outrageous prices to the highest bidders. Much of the existing vaccine went to private companies, grocery stores, chain pharmacies, big-box discount outlets, and other nonmedical establishments. Doctors' offices, hospitals, and long-term care facilities—where the drug was needed most—were left high and dry. That fall, U.S. Surgeon General David Satcher urged health care providers to "just say no" to healthy people who demanded a shot. The vaccine had become the "health equivalent of caviar," according to one reporter. As a member of the California Medical Association put it, "Right now in San Francisco, it's easier to buy heroin off the street than to get a flu shot from your doctor."

One of the strongest proponents of vaccine preparedness is Edwin Kilbourne, whose lab pioneered the development of high-yielding recombinant viruses used to make vaccines. Kilbourne first saw the human side of influenza up close in 1947, when a devastating epidemic hit Fort Monmouth, New Jersey, where he was serving at a military hospital. The vaccine that year completely failed, and beds spilled into the corridors. As the assistant chief of medicine, Kilbourne frantically tried to prevent secondary meningitis and to contain a streptococcal epidemic that followed on the heels of the flu. The experience was seared into his memory as he went on to study virology at the Rockefeller Institute. Today, Kilbourne believes scientists should be fabricating high yield reassortants for all 15 hemagglutinin subtypes, before a bizarre animal virus strikes anew. "You can talk about harnessing the Red Cross and the fire department," he says, "but in terms of specific final control of an epidemic, it's going to depend almost entirely on vaccine." The CDC and National Institutes of Health have carefully embarked on the process, with an H5N1 vaccine OK'd for human trials and an H9N2 precursor vaccine on the way.

Sadly, political issues still bedevil vaccine production. Who—the government or the drug companies—bears the financial risk if a pandemic fizzles out, or if a vaccine has unwelcome side effects? In a worldwide pandemic, political considerations will guide which nations get the vaccine. Within the United States, politics will decide who gets the first rationed supplies. "It's being very hotly debated as to who the priority groups should be," said Maine state epidemiologist Kathleen Gensheimer. "When we had our meeting in Maine, we literally almost had fistfights break out between the medical community and those on the front line, such as firefighters and the sheriffs. Because who really is considered critical?" Thefts and a black market are likely.

In pandemic times, many seemingly scientific issues will hinge on nonscientific considerations. For instance, what will be the stated purpose of vaccine use, to prevent high-risk people from dying or to prevent the epidemic from spreading? The answer dramatically changes public health policy. Halting the pandemic means first vaccinating children, who are notorious vectors of new flu strains and everything else that's going around, and healthy young adults. "The fires of the epidemic are fed by healthy susceptible school children, college students, and employed persons who have many daily contacts and who are more mobile," writes Paul Glezen, epidemiologist for the Influenza Research Center at the Baylor College of Medicine. Ed Kilbourne disagrees. "On ethical grounds, you really have to target the people at risk of death," he said. "Excess mortality, which is primarily in older subjects, is not just a matter of the old man's friend—pneumonia—taking away those about to climb into a coffin. With increasing health in older people, these are very active, productive people."

When he thinks back to 1997 in Hong Kong, Keiji Fukuda sees both the necessity and the shortfalls of preparedness. "On paper, it is dramatic. In real life, it's more dramatic," he says. "The reality is that, no matter how much we prepare, we will be unprepared. It's kind of like getting ready for a hurricane. You can batten down the house and you can make sure that the windows don't get blown out. But when you go through the hurricane, you're going to know that you went through it. When you look around, you'll think, 'Boy, there's an awful lot of devastation.'"

***

After Hong Kong's H5N1 outbreak, it became clear that the virus had struck more than the 18 documented patients, of whom 6 died. During the CDC's investigation that winter, blood samples were taken from more than 3,000 individuals. As expected, poultry workers had the highest risk of infection. Residents who passed through the city's live bird markets also had increased rates of antibodies. Most of these donors never knew they were infected and never felt ill.

So what to make of the fact that hundreds of people were silently infected? That question gnaws at and divides public health experts. When researchers place any large group of people under a magnifying glass, they can always find background abnormalities that never become clinically important, conditions such as "precancers" that never turn into malignancies, or "spinal abnormalities" in people who will never suffer back pain. In Hong Kong, did massive screening turn up lots of people who were subclinically infected with H5N1 but never would have become ill? Were these rates of apparent infection merely a screening artifact? Or, on the other hand, did they suggest that a dangerous H5N1 infection was acclimating itself to humans? Put another way, does the benign presence of H5N1 or H9N2 in people represent an everyday viral spark or the smoldering start of a pandemic? "Maybe this is similar to someone lighting matches and throwing them into a bunch of straw," says Fukuda. "The matches may blow out, but sooner or later, the straw is going to catch on fire."

If so, it means the December 1997 chicken slaughter in Hong Kong, radical as it appeared, was probably necessary. Skeptics said afterward that the virus had gone about as far as it could go. The CDC's Nancy Cox disagrees: Because the virus was so widely disseminated in the Hong Kong markets, and because there were already so many human infections, there were probably more to come. Scientists now know that the H5N1 virus had been undergoing rapid evolution during the incident, acquiring a number of amino acids in its internal proteins that correlate with growth in the human host. Indeed, in 2001, Yoshihiro Kawaoka discovered that a *single* chemical base change in the PB2 gene permitted the virus to infect people. If H5N1 had begun to spread in a big way among people, there is no telling how it would have further changed. "We don't know for sure that a pandemic would have occurred," Cox says. "But given the fact that this is a very adaptable virus, it's more likely that in fact it would have managed to reassort. Or it would have managed to adapt to its new host." In the wake of

H5N1 and H9N2, Robert Webster advocates an even more drastic measure: remove the 2,200 live poultry markets from Hong Kong.

***

Flu pandemic planning is a tough sell. It's hard to persuade government officials to get ready for an unforeseeable, if inevitable, event. During the H5N1 epidemic, the CDC was forced to ask retired flu experts to come back to work. When the scare died down, the influenza branch received only a modest increase in funding, most of it to improve surveillance in this country. Unlike diseases such as measles or hepatitis or even chronic fatigue syndrome, flu research does not enjoy congressionally mandated line-item funding. The U.S. pandemic preparedness plan laments that "Funds available for influenza-related research have declined steadily during the past decade."

In *The Forgotten Pandemic*, Alfred Crosby wrote: "The Deity, jaded with omnipotence, seems to have posed Himself a paradoxical problem: just how deadly a disease can I create that humans will barely notice? His answer to His challenge was influenza."

Ordinary flu is a trivial infection for most of us, the scapegoat for every passing fever and intestinal upset. But influenza is also a constantly emerging bug, which could set off a worldwide cull 50 years from now—or tomorrow. It springs up in startling places and moves in bewildering ways. Having declassified its genetic file, we still don't know what makes it deadly. Having eavesdropped on its avian haunts, we still don't know why bird strains pounce on humans. Scientists have searched for clues in the Alaskan tundra, in sultry Hong Kong markets, in dead soldiers' tissues, in laboratory animals—and they still don't know why the virus is so randomly vicious, where it will strike next, or how to stop it. They have circled and circled around the virus, yet it remains a mystery.

"However long I study this organism," says the CDC's Keiji Fukuda, "it may well be completely elusive to me in the end." When the next pandemic comes, we may be as spellbound and helpless as the villagers of Brevig Mission.

Chapter 6

Infection Unmasked

I n the United States, 70 percent of all deaths are due to chronic afflictions. Unlike the classic scourges of humankind, these illnesses usually strike insidiously. Until the past few years, the mechanisms of their mayhem were all but invisible. But today, a growing number of researchers claim that these disabling conditions may be caused by infection. If true, the repercussions are almost beyond imagining.

To get a sense of this rapidly shifting ground, consider the chronic illnesses not merely suspected but proven to result from infection. All are open-and-shut cases, a single agent provoking a single disease. The human papillomavirus, for instance, causes more than 90 percent of cervical cancer. The hepatitis B virus accounts for more than 60 percent of liver cancer. The hepatitis C virus claims 8,000 to 10,000 lives here each year from complications of chronic liver disease, a number

that could triple by 2010. The first identified retrovirus, HTLV-1, sets
the stage for adult T-cell lymphoma decades after the original infection. (Retroviruses, including HIV, are RNA viruses able to make DNA
that incorporates itself into the chromosome of an infected cell.) The
Epstein-Barr virus, a herpesvirus that triggers mononucleosis in teen-
agers and young adults, produces in people simultaneously infected
with malaria a cancer known as Burkitt's lymphoma, the leading cause
of childhood cancer deaths globally. Human herpesvirus 8 causes
Kaposi's sarcoma, a malignant complication of AIDS. *Helicobacter pylori*, a spiral-shaped bacterium, induces peptic ulcers. And a host of
foodborne pathogens trigger chronic and autoimmune diseases, such
as Guillain-Barré syndrome and reactive arthritis.

This may be merely the starting point. Just as the germ theory
cleared the way for a deluge of discoveries about the sources of acute
infections, today's changing ideas about the nature of infectious disease, coupled with new techniques in the lab, have led to a rush of
claims about the origins of chronic diseases. In a kind of medical
Klondike, some of these theories are based on wild hope and dubious
information, and will yield nothing. Other discoveries may change the
practice of medicine. Researchers have found the Epstein-Barr virus,
for example, in Hodgkin's disease patients and in aggressive breast
cancers. Multiple sclerosis (MS) acts suspiciously like an infection, with
its odd clustering across populations, its high antibody levels, and its
surges and retreats. Juvenile onset diabetes, also known as Type I diabetes, may arise when a Coxsackie B enterovirus elicits an immune
response that damages the pancreas. Crohn's disease, a chronic and
painful inflammation of the bowel, looks so much like an infection, the
physician for whom it was named remarked in the early 1900s that it
resembled an intestinal form of tuberculosis; today, one suspected
pathogen is *Mycobacterium paratuberculosis,* a cousin of the TB bacterium. Up to half of the asthma in this country, as well as a large portion
of juvenile rheumatoid arthritis, may be caused by *Mycoplasma
pneumoniae,* a free-living organism smaller than most bacteria. The
sexually transmitted organism *Ureaplasma urealyticum* appears to cause
meningitis, pneumonia, and even death in newborns—and it too may

be linked to childhood asthma. Patients with gallstones have far higher numbers of clostridia and eubacteria in their intestines than do healthy people. Kidney stones may be caused by nanobacteria, microbes so tiny they are dwarfed by viruses. A woman's chances of delivering a premature baby rise dramatically if she has bacterial gum disease; according to one estimate, curing periodontal infection in pregnant women would prevent 45,000 premature low weight births in the United States each year. Schizophrenia, which strikes about one million Americans and impels 10 percent of victims to take their own lives, may be remotely caused by viral infection just before or after birth, which may derail key neural connections in the developing brain; another theory ties the disease to a parasite carried by cats.

Fifty years ago, this list might have sounded preposterous. "People forget how new this history is. We've only known about infections for a little over a hundred years," says Julie Parsonnet, a physician and epidemiologist at Stanford University. "For the first ninety years, all we thought about was: There is this acute infectious agent and it causes this disease. So you get infected with *Vibrio cholerae* and you get cholera, you get infected with *Salmonella typhi* and you get typhoid. These are short-term diseases, they come and go, and the organism is gone. This has been most infectious disease research." Now the questions are different. Can sly viruses and bacteria set off more subtle damage? Do the organisms that live in us all the time, traditionally dismissed as "benign commensals," actually betray us? Is our old view of infection—the view stamped by the hallowed germ theory—just half of the picture?

Even if scientists link only some chronic conditions to viruses or bacteria, doctors would draw on vaccines and antibiotics to prevent and cure, rather than heavy-duty last-hope drugs or lifestyle therapies that may or may not slow symptoms. "One huge advantage of looking for infections is that you can do something about them," says Robert Yolken, director of the Stanley Neurovirology Laboratory. "In general, we can deal with infectious diseases much better than we can deal with almost any type of human disease. And it's easier to treat an infection than to get somebody to change their diet or their lifestyle."

So dramatic is this turning point in medicine, enthusiasts with an eye to history have already christened it the Second Golden Age of Bacteriology, Koch's Postulates Part II, and the New Germ Theory. In 1999, David Morens, a medical epidemiologist at the National Institute of Allergy and Infectious Diseases, convened a group of leaders in this emerging field. Before the meeting, he jotted down a long list of recently discovered infectious causes of disease and highly suggestive associations. "When a bureaucrat like me makes a list like that, it can really open your eyes. A large number of diseases that I was taught were chronic when I was in medical school have proven to be infectious." Morens realized that the quest for hidden agents behind chronic disease is no longer confined to some dreamy future. "It's happening," Morens says, "but it's not happening fast enough."

Trend Setters

The patron saint of the reformed church of infectious causation is Barry Marshall, the Australian gastroenterologist who in the early 1980s discovered the bacterial origins of peptic ulcers, one of the most societally inflected diseases around. The traditional litany of ulcer causes had included stress, smoking, alcohol, excess gastric acid, and genetic predisposition.

Like many in the pathogen discovery game, Marshall and his collaborator, pathologist J. Robin Warren, were unwittingly following in the footsteps of their predecessors. As early as 1892, reports in the medical literature had described a spiral-shaped bacterium that inhabited the stomach. But conventional wisdom then and for the next century stated that bacteria could never survive the sterile, acidic environs of the stomach. In the early 1950s, when a leading American gastroenterologist failed to find signs of infection in more than 1,000 gastric biopsies, the infection hypothesis was put to rest for 30 years. When analyzing biopsied tissue, pathologists set their microscopes at 250 power, since there was nothing to look for but cancer. Beginning in 1979, however, Warren cranked up his microscope to 500 power and noticed in gastritis specimens what looked like S-shaped bacteria. In

1981, he shared his discovery with Marshall, who was immediately fascinated. They wondered if they had stumbled across, in Marshall's words, "some interesting Australian phenomenon." By 1982, after a year of analyzing tissue from ulcer patients, Marshall, only 30 years old and still in training at Australia's Royal Perth Hospital, and Warren, the more seasoned physician to whom he was assigned, were convinced that the bacteria were living brazenly in a zone the medical texts had declared off-limits. "We didn't set out to find the cause of ulcers," Marshall says. "It was just an academic exercise."

Their attempts to grow the bacteria in various culture media failed again and again. That April, an outbreak of drug-resistant *Staphylococcus aureus* hit the hospital, and the microbiology lab was barraged with specimens. What transpired evokes shades of Fleming's serendipitous discovery of penicillin. Preoccupied with the staph outbreak, the lab's technicians left Marshall's and Warren's petri dishes to languish in the dark, humid incubator over the long Easter holiday—for five days instead of the usual two. It was enough time for a crop of translucent bumps, known as "water spray colonies," to grow. After the microbes were smeared on a slide, Marshall peered at them through a microscope and saw dozens of organisms—what he would call, in the title of his now classic 1983 *Lancet* paper, "Unidentified curved bacilli on gastric epithelium."

But did they cause ulcers? Marshall took biopsies from healthy people and from ulcer-afflicted patients and, without saying which were which, asked Robin Warren to look for bacteria. As it turned out, everyone with bacteria had an ulcer. And so strong was the correlation between bacteria and stomach inflammation—a prelude to ulcer—Warren "had to buy a new kind of calculator to work it out," says Marshall. "An eight noughts one." Put another way, the statistical chance that the association was random was one in one billion. People require bacteria to get inflammation, and inflammation to get ulcers. Marshall reasoned that any drug that removed inflammation would cure ulcers. In 1983, he began successfully treating patients with antibiotics and bismuth. That same year, at an infectious disease conference in Belgium, someone asked Marshall if he thought bacteria caused at least some ulcer disease. Marshall shot back that he believed bacteria caused *all* stomach ulcers. "It was annoying to have me standing up pontificating," he recalls. "With so many 'experts,' it was impossible to displace the dogma. Their agenda was to shut me up and get me out of gastroenterology and into general practice in the outback."

Unfortunately, Marshall couldn't produce the crowning proof. Try as he did, he couldn't induce ulcers by feeding the bacterium to piglets. So in June 1984, he threw out the animal model. As he later reported in the *Medical Journal of Australia,* "a 32-year-old man, a light smoker, and social drinker who had no known gastrointestinal disease or family history of peptic ulceration"—a superb test subject, in other words—"swallowed the growth from a flourishing three day culture of the isolate." The volunteer was Marshall himself. He had poured five milliliters of beef broth into a petri dish, swirled it around until all the colonies of bacteria dissolved into what looked like murky chicken soup, closed his eyes, threw back his head, and swallowed, gagging a few times. Five days later, Marshall began vomiting every morning— classic symptoms of gastritis. A photo of him taken that week shows a glum-looking young man with dark circles under his eyes. For seven mornings in a row he was sick, suffering the rest of the days and nights with a gnawing sensation somewhere between hunger and nausea.

Helicobacter pylori has since been blamed, not only for the seething inflammation of ulcers (all but the relatively small number of cases produced by high doses of nonsteroidal pain relievers), but also for virtually all stomach cancer cases worldwide. Marshall's antibiotic cure is widely accepted, and today ulcer treatment is as straightforward as that for strep throat or sinus infection. Researchers in chronic disease regard Marshall's work as totemic. Marshall himself believes the

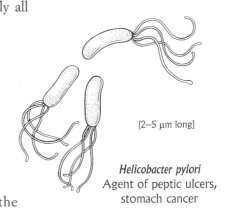

[2–5 μm long]

Helicobacter pylori
Agent of peptic ulcers,
stomach cancer

prospects for more breakthroughs are endless. "I think a lot of these chronic diseases are in the cards for being infectious," he says. "If you said to me, 'What do you think the chances are that antibiotics could protect you against cardiovascular disease?,' I'd say, 'Quite a good chance.'"

***

If Barry Marshall is the patron saint, Brent Muhlestein is a recent convert. Muhlestein is one of the leading researchers on the mechanisms of heart disease, the accumulation of fatty plaque in coronary arteries. An interventional cardiologist at Latter Day Saints Hospital in Salt Lake City, he used to perform angioplasties and then helplessly monitor his patients as their scooped-out vessel walls would block up again. In the hopes of predicting whose vessels would and would not close up, he saved the plaque he had cut out and analyzed the specimens' composition. Then, in 1993, Muhlestein picked up a copy of a publication he usually didn't peruse, the *Journal of Infectious Diseases.* Inside was an article that, as Muhlestein remembers, "sounded to me heretical." In heart vessel plaque taken from patients who had died mostly in automobile accidents, the authors had found *Chlamydia*-like organisms. "My mind said, 'What? Atherosclerotic plaque is sterile.' I couldn't believe it. But then my thought was, 'Well, I can prove them wrong by testing my ninety specimens.'" And so, he took his frozen plugs of plaque and looked for *Chlamydia pneumoniae*, which he knew simply as a respiratory pathogen. "Lo and behold," he reports, "we found that they were anything but wrong." Embedded in 79 percent of his plaque samples was the unsuspected bacterium.

Like Barry Marshall, Muhlestein had happened upon a well-trod historical path. In 1859, the pathologist Rudolf Virchow noted in coronary arteries "a stage of irritation preceding the fatty metamorphosis, comparable to the stage of swelling, cloudiness, and enlargement which we see in other inflamed parts." Virchow wrote that he was siding with what was, even then, considered "the old view in this matter" by holding "an inflammation of the inner arterial coat to be the starting point

of the so-called atheromatous degeneration." By the end of the nineteenth century, researchers could induce heart disease in animals by
injuring the coronary vessels and experimentally infecting the animals.
In his 1908 book *Modern Medicine,* William Osler listed "acute infections" among causes of arteriosclerosis. But during the first few decades of the twentieth century, the focus shifted away from germs and
toward the multifactorial or "black box" paradigm, which holds that
heart disease results from many different causes working in synergy,
from high cholesterol and hypertension to smoking and stress and
(what was later disproven) just being male.

In the 1970s, however, the infection theory again began to perco-
late as scientists pursued new viral and bacterial leads. When chickens
were experimentally infected with a herpesvirus, the birds developed
arterial lesions resembling those in human atherosclerosis. Then a 1985
Finnish study revealed that heart attack patients had higher levels of
antibodies to *Chlamydia pneumoniae* than did healthy subjects. All this
made researchers look twice at the mysterious decline in heart disease
that had commenced in the United States in the late 1960s—an unprecedented drop that continues to this day. Doctors knew the drop
wasn't due to a sudden society-wide passion for distance running and
macrobiotic cuisine. Could it have come from the widespread use of
broad-spectrum antibiotics, which inadvertently killed bacteria in the
heart?

Muhlestein decided to explore the bacterial hypothesis in his own
lab. He infected rabbits through the nose with *C. pneumoniae* and
found he could induce atherosclerosis. When he raised cholesterol levels in the animals' diets, the blockage worsened. Other studies have
found rising levels of antibodies to the bacterium during the months
before a heart attack. People infected with *C. pneumoniae* are also more
than four times as likely to suffer a first ischemic stroke—the kind
caused by blockage of a blood vessel—than are uninfected people. As
these discoveries piled up, the medical world had finally sanctioned
Barry Marshall's once-dissident belief that bacteria cause ulcers. "The
idea that another chronic inflammatory process like peptic ulcer disease was caused by a single infectious agent made me start to think,

'Well, maybe it's the same story,'" Muhlestein says. "Maybe there's just one infectious organism that's causing atherosclerosis that kills half the world. And if we kill that, maybe we can eliminate this terrible scourge." As Muhlestein and others would later come to realize, heart disease doesn't hew to such simple mechanisms.

Yet because it afflicts more than 50 million Americans and each year kills nearly one million, and because by 2020 it will be the leading cause of disability in the world, heart disease remains the prized trophy for microbe hunters. Are *Chlamydia pneunomiae* the causes of inflammation or merely innocent bystanders conveyed to the site by the immune cells in which they camp out? And what about the rest of the catalog of human misery—the chronic diseases and autoimmune disorders and psychiatric illnesses that continue to defy explanation? Will infections prove to be the long-sought answer or just a mirage?

Rules of Evidence

Until quite recently, the standard in infectious diseases for separating hope from hype, truth from speculation, have been guidelines known as Koch's postulates. The German country doctor Robert Koch, an enormously gifted lab man with an obsessive drive to answer the big questions in medicine, founded the field of bacteriology. To understand the importance of his postulates, you need to know a little about previous theories of infectious disease.

Ever since Aristotle's time, observers had believed that minute organisms arose spontaneously in decaying organic matter. One of Louis Pasteur's great achievements was his insistence that living organisms could not emerge without living precursors—that, as he put it, "the hypothesis of spontaneous generation is a chimera." This belief stemmed from his work in the 1850s and 1860s on yeasts and fermentation. In 1859, he declared: "Everything indicates that contagious diseases owe their existence to similar causes."

By demonstrating the life cycle of the anthrax bacterium, Robert Koch showed for the first time that a specific microorganism caused a specific infectious disease. In 1882, before the Physiological Society in

Berlin, he delivered an elegant paper on the etiology of tuberculosis. Paul Ehrlich, a future Nobel laureate then just in his twenties, proclaimed the occasion his "greatest scientific event." Koch's work ushered in the fruitful days of the germ theory, in which microbes suddenly seemed to unlock every puzzle. Researchers believed that yeasts could transmute into bacteria and that the hordes of bacteria normally found in tissue all contributed to disease. The enthusiasm engendered by an unsuspected world of disease agents was not unlike today's. Koch's sober postulates—a series of steps or procedures that had to be followed in order to prove that a microorganism causes a particular infectious disease (and actually the ideas of Koch's teacher, the German anatomist Jakob Henle)—was just the antidote to such irrational exuberance. Medicine needed techniques that would permit rigorously controlled experiments in which microscopic organisms could be isolated and handled.

Ironically, Koch himself never formally stated his namesake postulates, though the ideas shine through all his work. The German bacteriologist Friedrich Loeffler published them in a paper on diphtheria. Briefly, the postulates state that:

1. The organism must be present in diseased tissue.

2. The organism must be isolated and grown in pure culture outside the animal host.

3. The organism must be shown to induce the same disease when injected into a healthy animal.

Some reviewers have tacked on a fourth postulate: that the microbe must be isolated from the experimentally inoculated animal in pure culture and shown to be the same as the original agent.

Each infectious disease could be attributed only to a single species of microbe, not to every member of the microbial retinue found at the site. Microbe hunters had to isolate the bacterial culprit and prove its malignity. Those scientific hurdles were really a blessing. Between 1877 and 1897, investigators harvested a microbial bounty, including the agents of typhoid fever, leprosy, glanders, cholera, diphtheria, tetanus, gas gangrene, plague, dysentery, gonorrhea, cholera, common bacterial pneumonia, bacterial meningitis, and a main cause of food poisoning.

When he volunteered himself as a human guinea pig, Barry Marshall neatly fulfilled all of the postulates. Indeed, Koch's rules work beautifully whenever disease agents are easily cultured and transmitted, and when symptoms explode soon after infection.

But those same ground rules don't apply to more subtle medical mysteries. In chronic ailments, the links between infection and symptoms are obscure and covert. This is true for several reasons. First, suspect bacteria and viruses are more difficult to grow in the lab— sometimes impossible, as in the case of the hepatitis C virus. Even at the beginning of the twentieth century, scientists had complained of "less docile" organisms that didn't thrive in the usual culture media. Second, many human diseases suspected to have an infectious cause don't have animal equivalents, another roadblock in proving causation. Human studies are out, of course—it's unethical to subject people to experimental doses of agents that could cause chronic disease years later. Third, there can be yawning delays between infection and disease, so that by the time disease shows up, the agents that planted the original infection may have fled the scene. The seeds of some psychiatric diseases with onset in adulthood, for example, may be viral or bacterial infections that occurred just before or after birth, distorting events during a vulnerable point in the development of the brain.

Most discoveries about chronic infectious diseases over the past two decades—AIDS, adult T-cell leukemia, cervical and liver cancers— have for one reason or another failed to meet Koch's postulates. Those standards may even have deflected investigators from the new set of clues before them. Writes biologist Paul Ewald, "We need not, however, let the inability to meet the gold standard . . . lead us to be overly skeptical of assigning infectious disease causation based on silver or bronze standards." Or as microbiologist Hal Nash commented, "The nature of acute disease allowed for the development of 'rules' and 'postulates.' The nature of chronic disease is likely that there are no rules."

Rules or no rules, the evidence for infection in chronic diseases is too obvious to ignore. When an illness arises mostly in people whose immune systems are down, it's a tipoff to infection (take Kaposi's sarcoma, a complication of AIDS and transplant patients). When a disease improves with antibiotics (as does strep-induced rheumatic fever), it's a good bet it's infectious. The same is true when a disease occurs more often among people living in poverty (as does *H. pylori*-induced stomach cancer). Or when disease prevalence varies according to time of year (some data suggest that schizophrenia develops slightly more often in people born in the winter months—particularly February and March, the heart of the respiratory disease season). Or when the disease clusters geographically (multiple sclerosis rises in a steady gradient the farther one travels away from the equator, consistent with winter respiratory infections; an upsurge of MS from the early 1940s to the early 1970s on the Faroe Islands, in the North Atlantic, may have been sown by British troops during World War II).

And there's another clue—a kind of negative clue, if you will—that infections may play a central role in chronic diseases: scientists are beginning to realize how much they *don't* know about what's living in and on the human body. It is estimated that less than 1 percent of all bacterial species have been identified; only a tiny fraction of bacteria and viruses can be cultured with standard methods. Part of the problem is that we humans came late to the party. Prior to the last billion years, virtually all life on Earth was bacterial. From their very genesis, animals have probably carried indigenous bacteria. Even earthworms, which have been around for perhaps 800 million years, harbor such organisms. Having dominated the planet for three-quarters of its existence and having made themselves comfortable in just about every ecological niche—including *Homo sapiens*—microorganisms not only bombard our bodies but actually make up the fabric of our physical selves. Humans, you might say, evolved in order to haul around bacteria. A piece of doggerel from 1912 titled "On the Antiquity of Microbes" summed up the situation: "Adam had 'em."

Tricksters

How exactly do we become colonized by these tiny creatures? The fetus normally remains sterile as long as it resides in the uterus, protected by the placenta. Its first exposure to microorganisms usually takes place during passage through the birth canal. Immediately after delivery more organisms swarm in, some becoming permanently established in or on the baby's body. By the end of the first week or two, the infant has acquired most of the organisms found in the adult. And in most cases, that's lucky for us. The bowel alone contains hundreds of microbial species, some of which break down food into absorbable nutrients and may disarm potential carcinogens in the diet. In addition, intestinal microorganisms synthesize vitamins. Indigenous flora also stave off colonization by more virulent bacteria and stimulate the immune system. About 1 percent of the human genome consists of endogenous retroviruses—viruses that millions of years ago insinuated themselves into the human genome when they infected the progenitors of humans.

So abundant is our microbial haul that no one has even tried to draw up a complete bill of lading. Our bodies contain at least ten times more bacterial cells than human ones, making us walking petri dishes, and blurring the line between where microbes end and humans begin. Joshua Lederberg has described each human host and its complement of parasites as "a superorganism," with myriad genomes "yoked into a chimera of sorts." As Martin Blaser, chairman of the Department of Medicine at New York University, sees it, "The great dilemma of vertebrate biology is: How can we live with all these bacteria we're carrying? The short answer is that over millions of years nature has provided a way of regulating our immune response so that we're not so unresponsive that these guys kill us, and we're not so overresponsive that we destroy normal organs trying to fight these guys."

In other words, the one germ/one disease idea is too neat to encompass the teeming hordes that make up the human body. A supercomputer could barely keep up with the pathogens that are hitting us from without and from within, agents that potentially work in concert. "What may be causing chronic diseases," says Stanford University's Julie Parsonnet, "is the interplay between our responses to all those organisms simultaneously."

David Relman, a physician and microbiologist at Stanford, began to grasp the enormity of science's cluelessness when he set out to catalog bacteria in one of the most prolific bacterial niches: the subgingival crevice, which is the pocket between the gum and tooth. The abundance of action at this site had long been known. In a 1683 letter to the Royal Society, Anton van Leeuwenhoek, the Dutch lens grinder and jack-of-all trades, included drawings of bodies observed in scrapings from his own teeth. Peering through the microscope, he reported "many very little living animalcules, very prettily a-moving"—the first bacteria ever seen and described. Three centuries later, Relman took a sample from a healthy mouth and divided it in two. One half he gave to technicians in an academic hospital clinical microbiology lab, asking that they culture everything they could, separating all the organisms that appeared distinct and different. The other half Relman analyzed using molecular biology tools. While there was a fair amount of overlap in the two sets of results, Relman's approach uncovered almost ten times as many novel organisms as did traditional culturing. In one tiny sample, he found between 30 and 40 never-before-seen species. The statistical margin of error suggests that at least a dozen more were left undiscovered.

Relman concluded that the old idea of microorganisms either parasitically sapping the human body or benignly renting space doesn't capture reality. Some of the organisms that compose "endogenous flora"—our tiny corporeal cohabitants—could actually make us sick. Maybe it's not just barbarous pathogens from outside, ill-adapted to the human body, that give us grief, but also the quiet habitués that seem to biologically blend in and speak the same language. "There may be a whole lot more about what we carry around within our bodies, and what our bodies are in fact made up of, that could be factors in chronic disease," Relman says, "even in acute disease, for that matter."

How might these infinitesimal homebodies cook up disease while eluding detection? One strategy is known as persistence. Scientists used to think that viruses gave themselves away by telltale signs: they killed cells, they turned off protein synthesis, they replicated like mad. Under the microscope, virus-infected cells looked bloated and crummy until they finally fell apart. But viruses that cause persistent infection do not destroy cells. Rather, they can replicate continuously in so-called differentiated cells—those with special functions, such as the neurons that make neurotransmitters and synapses, or the endocrine cells that make hormones such as insulin. Such viral strategies "have been clearly established in animal models and are very likely to occur in humans," says Michael Oldstone, a virologist at the Scripps Research Institute in La Jolla. "What you wind up with are tissues or cells that look normal under a microscope but in fact contain infectious agents which are turning off the function of the cell. That would reflect, we think, the majority of human conditions that don't have any kind of etiology associated with them." Moreover, such diseases as MS, juvenile diabetes, and various psychiatric illnesses may be initiated by certain viral infections and promoted by the immune system's response to subsequent infections from unrelated agents, a kind of "multiple hit" operation that scientists haven't been able to pin down.

Microorganisms can also mimic components of the human body and thus spur autoimmune disease. A common denominator in these afflictions—from MS and rheumatoid arthritis to Crohn's disease and lupus—is chronic inflammation, a sign of possible infection. "One of the ways that infectious agents survive in humans is by making themselves look human, so that our body doesn't combat them," explains Julie Parsonnet. "But this system may go slightly awry, where the organisms look just enough off so that the body will respond to both the human cell and the infectious agent, thinking that they're both bad. The infectious agent may eventually be gone, but the human body will still think a foreigner is present because the human cells will be there." Autoimmune disease, adds David Relman, is the "consequence of this imperfect balance between the immune system's need for vigilance and its need for tolerance of structures that should not be attacked."

Yet exactly which infectious agents provoke autoimmunity, and how, is a mystery. Human immune defenses "are acting as if they would be reacting to a bug," says Thomas Quinn, a senior investigator at the National Institute of Allergy and Infectious Diseases. "But we've not been able to find the bug."

One scheme behind autoimmune disease is what scientists call "molecular mimicry." Organisms have evolved surface substances, called antigens, that closely resemble substances found in human tissue. Antigens are what raise specific antibody and T-cell (or T-lymphocyte) responses from the immune system. Most autoimmune diseases target one organ. The T cells that recognize an antigen on a Coxsackie B virus, for instance, sometimes also react with part of an enzyme in the cells of the pancreas that make insulin. When the T cells attack and destroy the pancreatic cells, diabetes results. Regardless of what specific viruses set off an autoimmune cascade, says Michael Oldstone, subsequent infection from an unrelated virus may exacerbate the damage.

One of the more surprising revelations of immunity gone awry is a childhood form of obsessive–compulsive disorder (OCD). Susan Swedo, a neuropsychiatrist at the National Institute of Mental Health, has found that infections with *Streptococcus* pyogenes, or Group A strep, in some children lead to a form of OCD sometimes accompanied by tics. The disorder may arise when proteins on the outside of strep bacteria stimulate the immune system to attack human brain tissue containing nearly identical proteins—again, the old trick of molecular mimicry. Swedo named the disorder PANDAS, for pediatric autoimmune neuropsychiatric disorders with strep. But clues to this unusual complication were known as early as the turn of the twentieth century. William Osler, the Canadian-born dean of American medicine, noticed that patients suffering from Sydenham's chorea—involuntary movements from neurologic complications of rheumatic fever, caused by streptococcus A—exhibited "a certain perseverativeness of behavior."

"In general," says Swedo, "other types of obsessive–compulsive disorders start slowly, gradually increase in severity, and if you ask somebody to tell you when it started, they're not able to do that. It's normally, 'Oh, I had it all my life,' or 'I think it was sometime in the first or second grade.' In these kids, the parents could say, 'It was the Tuesday before Thanksgiving.' It came on very quickly, very abruptly, and very severely." Fortunately, many children return to normal—but if they contract back to back strep infections, their symptoms can last a lifetime. Swedo estimates that about 10 percent of OCD in the United States stems from childhood strep infections—meaning that if 1 in 100 Americans have OCD, 1 in 1,000 have PANDAS. Swedo is now running an antibiotic trial to see if the drugs prevent second strep infections, and thus a lifelong disorder.

The immune system's attack against retroviruses embedded in the human genome may induce other psychiatric conditions. Pediatrician and virologist Robert Yolken has found that 30 percent of patients suffering first episodes of schizophrenia, and 7 percent with chronic schizophrenia, show signs that an ancient retrovirus in their brain is turned on; the molecular footprint didn't appear in the brains and cerebrospinal fluid of people who did not have schizophrenia. The virus—which can be activated by herpes simplex virus infections, as well as by hormones and immune cells—may or may not actually cause schizophrenia. To find out, Yolken plans studies in which he administers to these patients antiviral drugs in addition to antipsychotics, in the hope that their symptoms diminish. Yolken and psychiatrist E. Fuller Torrey have also proposed that some cases of schizophrenia and manic depression may result when the fetal brain is infected through the mother with the parasite *Toxoplasma gondii*, an organism transmitted by cat feces.

If true, such theories may cast a new light on older medical literature. Many nineteenth-century doctors surmised that schizophrenia could be an infectious disorder. "It is certain that there are years when . . . insanity seems suddenly to extend to a great number of individuals," wrote the French neurologist Jean E. Esquirol. In 1845, when Esquirol observed that "mental alienation is epidemic," he meant it literally.

Evolutionary Thoughts

Amherst College's Paul Ewald offers an even more provocative twist on evolution. He contends that many viruses and bacteria have evolved in such a way that it's in their interest to cause latent, languishing disease—to be "cryptic," as he puts it, rather than conspicuous. Ewald is not a virologist or molecular biologist or physician; his ideas are rooted in the more speculative realm of evolutionary biology. Nevertheless, a number of researchers and theorists at the front lines of pathogen discovery have been struck by the resonance between their work and Ewald's. As one virologist put it, Ewald's theories could raise "red flags" that point scientists to certain infectious agents behind chronic disease.

Like a good evolutionary biologist, Ewald assumes at the outset that the raison d'être of microorganisms and humans alike is to pass on their genes. He goes on to divide human disease into three categories: genetic, environmental, and infectious. Deleterious genes—those that harm survival or reproduction by causing disease—are weeded out by natural selection. Environmentally caused illnesses—such as lung cancer from smoking cigarettes or certain cancers from Agent Orange—are limited both in time and space. And so, Ewald reasons, if a serious and widespread disease is not genetic—that is, if it is too common to have sprung up by random mutation and too ruinous to have survived the culling process of natural selection—and if it is not environmental, it must therefore be infectious. He applies this to today's most damaging diseases: atherosclerosis, stroke, many if not most cancers, brain disorders, and autoimmune conditions.

Even some diseases we put under the "genetic" heading—such as sickle-cell anemia or cystic fibrosis—are indirectly forged by infectious threats, Ewald theorizes. People with two copies of the sickle cell anemia gene, for instance, suffer the painful blood disorder—but those with one copy of the sickle-cell anemia gene and one normal gene are not only spared sickle-cell anemia but are also less susceptible to the most serious type of malaria. Likewise, healthy people who have one cystic fibrosis gene and one normal gene appear more resistant to typhoid fever. Seen through the lens of evolutionary biology, genetic mutations that occasionally gum up a human machine fine-tuned over millennia by natural selection may actually benefit us by thwarting the equally fine-tuned mechanisms of the microorganisms besieging us.

According to Ewald, if one considered only their effects on human health, scourges such as smallpox, TB, malaria, typhoid, bubonic plague, yellow fever, and cholera could *only* have been infectious. By the same token, it should have been no surprise that the rising rate of female infertility in the 1970s turned out to be primarily caused by a sexually transmitted agent, *Chlamydia trachomatis*; more liberal attitudes about sex, coupled with the Pill and other nonbarrier methods of birth control, gave the bacterium a free ride. Even the common diseases of older individuals that don't impair reproduction—diseases such as atherosclerosis and various forms of dementia—must by a process of elimination be infectious, Ewald contends; since identical twins don't automatically suffer the same condition, genes can't explain it all, and since risk factors like diet and smoking fail to predict many cases, neither is environment the last word.

Bucking the conventional wisdom that disease-causing organisms evolve toward benign coexistence with humans, Ewald early in his career proposed that infectious agents can become either more or less virulent over time, depending on how durable they are and how they are transmitted. Drawing on this earlier work, Ewald makes a new prediction: the pathogens behind many chronic diseases are probably transmitted either through sex or through some other intimate contact, such as kissing or hugging. After all, these agents—unlike, say, flu or Ebola viruses—depend on mobile humans to spread. And because they rely on people to get around, it's not in their interest to quickly dispatch their hosts. According to Ewald, sexually transmitted pathogens "have to have tricks up their sleeves for avoiding the immune system." They must be infectious for months and transmissible after a long period of infection. No wonder, he says, that today's leading cast of characters in chronic diseases—the Epstein-Barr virus, HHV8, the human papillomavirus, the hepatitis B virus, as well as HIV—are all spread through intimate contact. Those cunning pathogens are most likely to escape the immune system's security forces. "A tremendously disproportionate number of agents responsible for causing chronic diseases," says Ewald, "will be STPs—sexually transmitted pathogens."

Delving into medical history, Ewald has concluded that the more cryptic a disease's chain of transmission, the more reluctant medicine has been to acknowledge an infectious cause (though there are exceptions—dengue and cholera, for instance, were widely speculated to be infectious, though their chains of transmission were elusive). "We're doing the same thing with cancer, heart disease, diabetes," he says, "that people were doing in the 1840s with yellow fever and gonorrhea and malaria and diarrhcal diseases that were transmitted by water." Moreover, the triumph of the germ theory led to an arbitrary dichotomy between acute and chronic diseases. Back then, newly found infectious diseases had a defined incubation period and fever as a prominent symptom. Chronic afflictions such as heart disease and cancer were often set aside as degenerative, age-related, and by definition noninfectious. Doctors pictured their complex origins as a web of causation, a crowd of risk factors—in drier language, as "multifactorial."

But to Ewald, "multifactorial" is a weasel word. He contends that the beauty of scientific medicine is in declaring causal hypotheses, then proving or disproving those claims. "Risk factors, in and of themselves, don't tell you anything about causation," he says. "The real answers in nature do involve causes." Ewald's stark and simple definition of an infectious disease is as follows: a disease that would be eliminated if its infectious agent were eliminated. And he applies it to acute and chronic diseases alike.

The only way to get to the real answers—and here, Ewald and others in the field agree—is to perfect new methods of ferreting out disease-causing organisms, to catch them in their tracks. "New tools will allow us to make cryptic infectious agents less cryptic," says Ewald—words that could easily have been the rallying cry of van Leeuwenhoek or Koch.

Higher Resolution

The father of bacteriology was also the father of invention. To abet his rural medical practice, Robert Koch built a state-of-the-art microbiology lab. An ingenious innovator, he adapted the light microscope to better magnify bacteria, contrived fixing and staining methods to better see them, developed solid nutrient media techniques for growing them, and was the first to publish their photomicrographic portraits. As much as anything, it was Koch's technical achievements that established the germ theory. Today, the petri dish and tissue staining have given way to nucleic acid amplification and gene chips. These molecular biology techniques go beyond appearances to plumb pathogens' genetic makeup, an infinitely more precise way to distinguish one agent from another. "It's like the difference between looking at a computer screen that only has three hundred lines versus twelve hundred," says Stanford University microbiologist Stanley Falkow. "You can see things more sharply and clearly." Though the leads for today's explorations have been known for decades, adds Stanford's David Relman, "They've been buried within the pathology literature and some of the clinical literature. They could only remain theories. There weren't the means of pursuing these ideas experimentally—until now."

Relman helped carve out the nascent field of pathogen discovery. His interest was first piqued in 1989 at a weekly clinical conference. Doctors had noticed hordes of curved rod bacilli in a disease called bacillary angiomatosis, a complication that causes skin lesions and systemic problems in AIDS patients and other immunocompromised people—but the bacterium refused to replicate in culture. Relman tackled the problem by throwing out Koch's postulates. Instead, he employed a technique called consensus polymerase chain reaction (PCR). Using snatches of genetic sequences common to all bacteria as a kind of bait, he snagged and then copied a small piece of the unknown bacterium's genome. Comparing that piece with the genes of known agents, Relman stumbled upon a surprise: the bacillary angiomatosis bacterium was closely related to *Bartonella quintana*, the agent of World War I trench fever. When Relman further refined the identification, he came up with another surprise: bacillary angiomatosis came from the same bacterium—*Bartonella henselae*—that causes cat scratch disease, a mild, flu-like illness in humans. In the ever-growing annals of emerging infections, here was another case of an old bug showing up in a new host. Relman followed up his tour de force by using the same genetic angling techniques to find the bacterium that causes Whipple's disease, a puzzling wasting syndrome long suspected to be infectious. But other scientists were reluctant to accept his findings. "We simply had a DNA sequence out of these tissues—we didn't have a living organism," Relman recalls. "For a lot of people, that didn't feel quite right." Today, however, Relman's experiments are considered critical turning points.

Another technique—representational difference analysis, or RDA—selectively amplifies rare genetic fragments that show up in diseased tissue but not in otherwise matched normal tissue. This was how Yuan Chang, a pathologist at Columbia University, discovered the infection that causes Kaposi's sarcoma, or KS.

Before the era of AIDS, Kaposi's sarcoma, a rare malignancy that causes purple blotches on the skin, was known mainly as a benign cancer in elderly men of the Mediterranean and the Middle East, and in men of Jewish descent (the so-called classic form of the disease); as endemic disease in equatorial Africa; and as an occasional side effect in immunosuppressed organ transplant patients. But in 1981, the cancer entered the spotlight when it alerted doctors to the AIDS epidemic. Early in the epidemic, the risk of developing KS was 10,000 times higher among HIV-positive patients than in the general population. (In fact, one of the original names for AIDS at the CDC was Kaposi Sarcoma–Opportunistic Infection.)

For years afterward, doctors assumed these KS lesions resulted from HIV's own effects on endothelial cells. But by the early 1990s, Chang and her husband, epidemiologist Patrick Moore, had other ideas. Certain peculiarities hinted that KS sprang from an infection. Epidemiologically, it seemed to spread like one. While HIV-positive gay men often developed KS, hemophiliacs with HIV had very low rates of the cancer, suggesting that it might be sexually transmitted. The classic and endemic forms of the disease followed a patchy geographic pattern, another clue to infection.

Moore, who had researched arboviruses for the CDC, worked for the New York City health department. Chang, a pathologist specializing in brain cancer, had just set up, on a shoestring budget, a molecular diagnostic lab at Columbia. As they were pondering the problem in 1993, scientists at Cold Spring Harbor Laboratory published the first paper on representational difference analysis. "It was a really beautiful technique," says Chang. "We just had to try it."

Theirs was a magnificent obsession. To carry out their experiment, they each worked their regular jobs and then holed up until late at night in Chang's lab. With no padding in her budget, Chang borrowed reagents from other laboratories. "We were working on average fourteen to sixteen hours a day," Patrick Moore says, "getting three or four hours' sleep in between the chores. That went on seven days a week for weeks and weeks." Adds Chang, "We knew all the gypsy cab drivers around Columbia." Most of their time was spent trying to prove themselves wrong—"because," explains Moore, "you can fall in love with a hypothesis and it can really trick you."

Using RDA on KS lesions, Chang found two small viral fragments. She and her husband later showed that the virus was present not only in AIDS patients with Kaposi's sarcoma but in all patients who had KS. The scientists eventually devised a blood test and a genetic test that picks up the virus and predicts disease before symptoms ever arise. Searches through a gene library revealed that the virus resembled the Epstein-Barr herpesvirus, one of just a handful of viruses known to cause tumors. Eventually, the 34-year-old Chang proved the cancer was caused by a never-before-seen herpesvirus, which now variously goes by the names human herpesvirus 8 (HHV8) and Kaposi's sarcoma herpesvirus (KSHV). The Chang and Moore theory may have been bolstered by a 2001 report from University of Michigan scientists, who found that another herpesvirus, the Epstein-Barr virus, disables a brake on cell migration and thus permits cells to spread and metastasize.

Key discoveries so far about the biology of cancer have all come from learning how tumor viruses convert normal cells into cancer cells. In 1909, the Rockefeller Institute's Peyton Rous discovered the first known tumor virus in chickens—though it was not until 1966, when Rous was in his mid-eighties, that he finally was awarded the Nobel Prize in medicine and physiology. Indeed, five recent Nobel prizes awarded for cancer research have gone to scientists studying tumor viruses (in 1926, the Danish pathologist Johannes Fibiger won for his discovery of a parasitic cause of stomach cancer in rats, which was used as a method of inducing cancer). What makes these viruses so illuminating is that they turn on and off many of the same genes that, when spontaneously mutated, cause cells to turn malignant. In a somewhat controversial theory, Chang and Moore now propose that the same mechanisms designed by the immune system to control viral infections are intertwined with those designed to hem in tumors. "When the virus is preventing an anti-viral response from occurring—to save its own life, to maintain itself, and to be transmitted—it accidentally can trigger a cell into a transformed or cancerous phenotype," says Moore, who now teaches at Columbia. This implies that if researchers want to find out how all cancer starts, they should study the cellular checkpoints erected to halt viral infections.

Chang and Moore's work, now embraced by their colleagues, was the first high-profile discovery of a new agent using methods that left cell and tissue culturing in the dust. "It was very high risk. It was a new technique," says Chang. "A group of people had been looking for a long time and were naysaying the idea of an infectious agent." When her first paper on the virus was published in *Science* in 1994, an accompanying editorial cast some doubts on the finding. "It's really good work, and it could be a substantial advance," commented the National Cancer Institute's Robert Gallo, one of the co-discoverers of the AIDS virus. "But I have major questions." Chang vividly remembers reading those words. "It was like the 600 pound gorilla stomping on a bug." Nevertheless, she has embarked on more searches. "I feel very strongly that other chronic diseases and cancers are caused by infectious agents that have not been identified yet, or where the association has not yet been made."

An even newer generation of technologies—known as DNA chips or DNA microarrays—can register tens of thousands of genetic sequences in viruses or bacteria at once. These chips are rectangles of glass or silicon roughly the size of a microscope slide that contain thousands of known gene sequences, to which complementary sequences from a patient's tissue will bind. David Relman and others suggest that such chips could be used to regularly survey normal bacteria in the mouth or bowel or on the skin—creating a detailed portrait of healthy tissue. Doctors could then compare at a molecular level this inventory of "normal" body flora against disease aberrations as they unfold, when they are presumably more treatable. Other chips can simultaneously register the turning on and off of thousands of human genes in response to infectious agents. Indeed, scientists envision a not-too-distant future when they will actually be able to diagnose disease by analyzing the unique pattern of expression as genes react to infections. As David Relman and colleague Craig Cummings have written, "We are on the verge of being able to listen to the two-way conversation between pathogen and host through devices of immense power."

Warning Lights

As these new techniques suggest, the easy part will be finding new bacteria and viruses, or cornering them in surprising hideouts. The hard part will be proving in a scientific court of law that they really do cause disease and are not merely hanging around the scene of the crime. "That's the downside to all this," says the National Institute of Allergy and Infectious Diseases' Thomas Quinn. "If you rush to publish without really good evidence and support, it discredits the field." So many agents have been floated as the "germs" behind so many conditions—even obesity—headline writers have had a field day. But unlike *H. pylori* and ulcers, further research often paints a more ambiguous picture.

The search for the genesis of heart disease illustrates these problems. Researchers know that patients with atherosclerosis have higher levels of antibodies against not only *C. pneumoniae* but also *H. pylori*, cytomegalovirus, herpes simplex viruses, *Mycoplasma pneumoniae,* and the hepatitis A virus. Brent Muhlestein, an early believer in the *Chlamydia* hypothesis, now concedes, along with many cardiologists, that the issues are messy. "Maybe coronary heart disease is not going to be due to a simple single infectious agent," he says. "It would more likely be associated with a variety of infectious agents and probably with other risk factors. Atherosclerosis is a tremendously multifactorial disease. We may never find one single unifying cause. Although one can still dream of such a thing." Supporting this idea, a 2001 study by Austrian and Italian researchers found that chronic infections of all kinds— from gum disease and bronchitis to sinus and urinary tract infections— triple the risk of atherosclerosis.

Other heart specialists go so far as to cast doubt on the whole infection theory. While atherosclerosis is clearly an inflammatory disorder, they say, the inflammation could spring from lots of things besides infection—the oxidation of "bad" LDL cholesterol, toxins from cigarettes, a misdirected autoimmune reaction. That people who have had heart attacks in the past exhibit higher levels of antibodies to certain respiratory organisms could just mean that they were sicker to begin with—that their lungs were weaker and they were more prone to infection. When Paul Ridker, a heart researcher at Boston's Brigham and Women's Hospital, published a study in 1997 linking heart disease to inflammation, *Newsweek* blared on a lurid red-and-yellow cover a story about "The Hidden Causes of Heart Attack," including bacteria—a theory that Ridker himself all but dismisses. "My experience with the *Newsweek* cover was: It's so easy for the lay public to equate inflammation, which they don't understand, to infection, which is so intuitive," he says. Many of the tantalizing reports linking microorganisms and heart disease that do hit the daily news cycle "would not pass scientific muster," Ridker adds, because they narrowly focus on one experiment and do not take into account the massive backlog of contradictory studies.

Even if the infection hypothesis does turn out to be true, treatment would be tricky. Vaccines targeted against the organisms that cause heart disease could actually aggravate any autoimmune process triggered by the pathogen, since the vaccine would provoke the immune system in the same way as the infectious agent. Antibiotics present problems of their own. For one thing, since these drugs don't kill specific bacteria but rather a broad range of microbes, doctors may not know what particular pathogen may be causing heart disease. And, of course, as we've seen in Chapter 4, long-term use of broad-spectrum antibiotics forces the evolution of bacterial strains that can withstand the drug. "If I give 10,000 people an antibiotic for a year or two years or five years or ten years, the one thing I can guarantee you is they will start developing resistant organisms," says Ridker. "Until you have definitive evidence that you are doing some benefit, you damn well better do no harm." To help settle these questions, scientists have enrolled more than 10,000 people in prospective clinical trials that will try to determine, as a starting point, whether antibiotics do reduce heart disease. Similarly, in 2001, Australian researchers began a clinical trial in which Crohn's disease patients were treated with antibiotics.

The quest to eradicate chronic diseases may also bring unanticipated problems. Take the case of *Helicobacter pylori*. Barry Marshall's curative treatment for ulcers, together with the wider use of broad-spectrum antibiotics, has killed off loads of *H. pylori* among residents of developed nations. That's good news if you're prone to ulcers. But some scientists believe that the bacterium, while causing ulcers, also protects against gastric reflux by producing an enzyme that reduces acidity (and makes it possible for *H. pylori* to survive in the stomach). So as ulcers decrease, reflux and its complications may rise. In the United States, adenocarcinoma of the esophagus is the fastest increasing cancer, and reflux is a key risk factor. "For the first time in human history, you have large numbers of people who are reaching their fourth, fifth, sixth decade of life without *Helicobacter* who still have full acid-producing mechanisms, and that's leading to reflux and its sequelae," says New York University's Martin Blaser. "One is a known risk that is decreasing, and the other is an unknown risk that's increasing and we don't know where it's going to end." Blaser is convinced that *H. pylori* evolved in our species not so much to protect against reflux as against infant diarrhea. "We have evolved with a certain microbial flora over the last millions of years," he says. "*Helicobacter* was normal flora, and the twentieth century is the aberration. In brief, don't mess with Mother Nature."

Perhaps the biggest paradox about the emerging field of pathogen discovery is that, despite its promise, funders are wary. "When you do fishing expedition research, the hits are few and the misses are many. Science wants to be safe," explains the National Institute of Allergy and Infectious Diseases' David Morens. Mainstream institutions favor "hypothesis-driven" rather than "hypothesis-generating" research. In the former, a scientist starts with a supposition and conducts the experiment to prove or disprove the idea; whatever the results, at least in the end there's a paper to write up, something to show for the work. But in hypothesis-generating research—the engine behind today's search for new pathogens—the scientist inches forward by hunch and intuition, gathering clues, speculating on what they mean. "Nobody's funding hypothesis-generating research," says Morens. "Within science, that's considered the lowest level of research—out of the primeval ooze."

But anyone who searches for new pathogens must wander over that muddy, uncharted terrain. Yuan Chang took the risk—and it paid off. But while "KS was great at perking people's ears up and alerting them to the fact that it can be done," she says, "most associations that will be found in the future will be a little less clear than the epidemiology for KS. The only way to look for new infectious agents or a new association between an infectious agent and a disease is to have a lab devoted full-time to that."

Scientists who go out on a limb often complain that the researchers who volunteer to sit on the National Institutes of Health study sections that approve or reject research proposals are not sophisticated or imaginative enough to assess more daring scientific visions. That bureaucratic process, they say, leaches innovation from the field. In pathogen discovery and in other realms, the answer may be to financially support the best minds for long stints, letting them go wherever their ideas take them.

The science establishment's reluctance to pursue unproven ideas is a hurdle medical oncologist Beatriz G.-T. Pogo knows well. At the Mt. Sinai School of Medicine in New York, Pogo has tested more than 1,000 breast tumor samples from the operating room and has found that more than a third have genetic swatches nearly identical to mouse mammary tumor virus, a slow-growing retrovirus that is transmitted through mother's milk in mice and causes cancer in female progeny. In these human tissues, the virus appears to be exogenous—i.e., it is not part of the women's own genome but rather comes from outside—and it virtually is never found in normal breast tissue. Pogo doesn't know how the virus is transmitted, nor exactly how it infects cells. "I am the first one to acknowledge there are many holes in this story," says Pogo. But the very suggestion that some breast cancer could be caused by infection raises the hackles of the medical community. Pogo's hypothesis must compete with theories about genetics and hormone exposure—ideas that have a lot of good data and adherents behind them. Before she embarked on this research, in the early 1990s, Pogo specialized in studying pox viruses, another group known to cause cancer, and she was well sponsored. Not so once she turned to potential infectious causes of breast cancer. "All my life, prior to this subject, I have been funded by NIH," she lamented. "But this subject has been considered high risk."

Barry Marshall, who broke the ground that others are now sowing, understands these feelings. Had he not discovered in the 1980s that bacteria cause ulcers, he says, today he might not get funding to look. Proceeding on intuition, Marshall says, "is a luxury that not many researchers have. It was always said to me that if you want to get research funding you better make sure that you've got the experiment half done. You have to prove it works before they'll fund you to test it out." And it helps to be something of an outsider. "The people who have got a stake in the old technology are never the ones to embrace the new technology. It's always someone a bit on the periphery—who hasn't got anything to gain by the status quo—who is interested in changing it."

Future Perfect

Even if only some of these hunches prove right, the payoff could be enormous. Doctors will be able to define chronic diseases by the pathogens that cause them and single out who is at risk. Indeed, they may someday be able to predict who is likely to develop a chronic disease even before it starts. If antivirals or antibiotics stop the problem, the next step would be immunization to prevent it. This is what Robert Yolken calls Koch's Fifth Postulate: "The true test will be removing these organisms and then seeing what happens to the disease. In Koch's time, they didn't have this luxury."

"If we were to eliminate all these infectious influences, maybe people would be quite healthy," says Paul Ewald. "They wouldn't have arthritis, they wouldn't have atherosclerosis, they wouldn't have Alzheimer's, they wouldn't have much of any cancer. If you take out all of those things that look like they might be caused by infection, maybe you would end up with people who are quite vigorous. Maybe at the age of ninety, one hundred, one hundred and ten, everything would just fall apart. We just don't know what real encoded senescence is."

But others are more reserved in their enthusiasm. "We've gone through these cycles before, where everyone thinks science is going to cure and solve everything. And it just doesn't," says Stanford's Julie Parsonnet. Two decades after scientists discovered the AIDS virus, for instance, there is still no reliable drug cure or vaccine. Unlike the years just after the germ theory was accepted, when acute infections fell like dominoes, chronic diseases may stand their ground for a long while.

Chapter 7

Bioterror

n the taxonomy of human warfare, artillery that pounds, pierces, cuts, blasts, or burns—even in the unexpected form of a Boeing 767—is deemed conventional. However horrific the events of September 11, 2001, what is considered unconventional, even beyond the pale, is attack via disease—which came just a few weeks later.

The autumn 2001 wave of anthrax cases—spread by spores that infiltrated post office sorting machines and mail storage areas, then government and executive offices and private homes, and eventually the lungs and skin and blood of defenseless victims—illustrated a danger about which terrorism experts had long warned: to those inclined to use them, viruses and bacteria make ingenious instruments of destruction. Unlike nuclear or chemical weapons, they self-propagate and adapt. They jump continents with ease. And a small amount can cause vast damage; after World War II, for instance, the U.S. Army con-cocted a botulinum toxin so potent that a pound, if expertly dispersed, could kill a billion people.

Chemical weapons are poisonous substances that are inorganic and manmade. Biowarfare agents, by contrast, are living organisms, or tox-ins secreted by living organisms, that can be used against people, ani-mals, or crops. When TNT or chemical bombs explode, emergency crews rush to the scene. In a bioterrorist assault, health officials—as we have so dismayingly witnessed—may not notice until too late. Be-cause pathogens require an incubation period to multiply in the body before they trigger symptoms, the first sign could occur days or weeks after the attack. By that time, the trail would be cold, and it might be impossible to determine if the epidemic was a freak event or a mali-cious act of aggression.

Ironically, this conundrum—the delayed onset of an epidemic and the opacity of its cause—created one of the jewels of modern public health. In 1951, the CDC's first crack team of disease detectives—its Epidemic Intelligence Service, or EIS, officers—hit the road. Twenty-two young doctors and a sanitary engineer trained in a public health boot camp, readying themselves to solve any strange or unaccountable outbreak. At the height of the Cold War, the impetus behind forming the team was the prospect of biowarfare, as the very image of an "intel-ligence" corps conducting disease "surveillance" suggests. "[A]ny plan of defense against biological warfare sabotage required trained epide-miologists," wrote Alexander Langmuir, the EIS founder and legend-ary chief of the agency's epidemiology branch, "alert to the possibili-ties and available for call at a moment's notice anywhere in the country."

Until September 2001, the public knew the CDC's gumshoes for their peacetime achievements—helping figure out Legionnaires' dis-ease, the Jack in the Box *E. coli* outbreak, the aspirin/Reye's syndrome link, hantavirus pulmonary syndrome, and countless other mysteries. Each year, EIS officers conduct between 800 and 1,000 such field in-vestigations. But on September 11 of that year—when terrorist-hi-jacked commercial airliners crashed into the World Trade Center Towers in New York City, into the Pentagon, and into a field in Pennsylvania—the EIS's original mission was dramatically revived. Within hours of the World Trade Center attacks, EIS officers fanned out to hospitals throughout the city, looking for evidence of novel disease syndromes. The CDC issued a nationwide alert to doctors and laboratories to watch for unusual symptoms and pathogens. The Department of Health and Human Services authorized shipments totaling 50 tons from the never-before-tapped National Pharmaceutical Stockpile, a massive cargo of antibiotics, ventilators, and other emergency supplies created specifically to counteract a chemical or bioweapons attack. In early October, when a Florida man became the first American in a quarter century to succumb to inhalational anthrax—the first clue to the organized dissemination of anthrax spores that soon preoccupied the country—the CDC immediately chartered a jet and sent 15 investigators, including EIS officers, to comb the area for signs of bioterrorism.

Before these cataclysmic events, the warnings that such terrorism was imminent had sounded at times like a five-alarm mantra: "It's not if," the chorus went, "but when." Their Cassandra predictions came true sooner than most had expected.

The New Line-Up

Bioterrorism agents kill by suffocating pneumonia, septic shock, massive bleeding, or paralysis. At the CDC, the worst of the worst are known as Category A agents: unusually deadly organisms readily spread in the environment or transmitted from person to person, that could trigger public panic and social upheaval, and that require special public health precautions. Topping the list are the agents that cause anthrax, smallpox, and plague.

After October 2001, the anthrax bacterium—a marquee agent of biowarfare, oft-threatened in the United States but never employed—suddenly became a familiar and credible danger. That October was a turning point in another way as well: while inhalational anthrax had always been considered a death sentence, aggressive medical intervention was shown to save at least some patients with the dreaded pulmonary form of the disease. Anthrax begins mildly enough, with fever, fatigue, and sometimes a cough. What often follows are a few cruelly deceptive days of respite. All the while, toxins released in the bloodstream by bacteria are destroying cells and causing fluid to accumulate in tissues. Leaky blood vessels cause blood pressure to plummet and organs to fail. Then comes the fatal blow: abruptly labored breathing followed, within a day or two, by shock leading to death.

The disease does not spread person to person, but by bacterial spores—tiny, hard-coated spheres that can survive for decades in the soil. This hardiness has been demonstrated every time workers have tried to clean out anthrax spores from the environment. In 1940, the British detonated anthrax explosives on Gruinard Island, a heath-covered outcrop off Scotland; four decades later, the spores were still viable. Crews had to deluge the island with a cocktail of 280 tons of formaldehyde mixed with 2,000 tons of sea water. In the 1970s, demolition experts tackled the spore-ridden Arms Textile Company building, on the Merrimack River in Manchester, New Hampshire. After the building was decontaminated with formaldehyde, anthrax spores survived in dust that had become embedded in the structure's cracks. Eventually, workers had to disassemble the plant and incinerate it brick by brick and board by board, then treat what remained with chlorine and bury the waste. It's no wonder that soon after the 2001 terrorist attacks, the United States and Uzbekistan signed an agreement to remove anthrax bacteria from a remote island in the Aral Sea, where the Soviet Union in 1988 had dumped hundreds of tons of weaponized spores from its biowarfare program—spores that remained not only virulently alive but vulnerable to theft.

The very properties that make anthrax a good bioweapon—its stability and lethality and ease of cultivation—recommended the bacterium to be the first "germ" that Robert Koch employed to prove the germ theory. The bacillus changes into spores when faced with lack of nutrients or water, or when subjected to chemical shock. It morphs back to its rapidly multiplying vegetative form when conditions are more palmy—for instance, in the nutrient-rich blood or tissue of the lungs. For this quirk of biology alone, anthrax is a perfect terrorist tool. As Ed Regis writes in *The Biology of Doom*, "Sporulation . . . was God's gift to germ warfare."

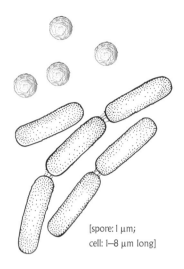

[spore: 1 μm; cell: 1–8 μm long]

Bacillus anthracis
(spores and vegetative cells)
Agent of anthrax

But anthrax possesses other features that make it a desirable weapon. The bacterium lives in soils worldwide and in its natural hosts, grazing livestock, thus making it widely available to anyone with nefarious motives. In the United States, anthrax zones closely parallel the nineteenth-century cattle drives from Texas to Montana. Back then, anthrax was primarily a livestock disease, as it is today; in 2001, an epidemic of anthrax killed cattle and bison in the upper Midwest and Canada. Grazing animals become infected by eating and inhaling large quantities of spores in contaminated dirt. In humans, most anthrax infections erupt on the skin. Indeed, the word anthrax comes from the Greek word for coal, referring to the infection's black cutaneous lesions. In the early twentieth century, common sources of cutaneous anthrax were imported animal hides and shaving brushes made from contaminated horse hair. Another form of the disease, gastrointestinal anthrax, is contracted by eating tainted meat.

Inhalational anthrax is the rarest form of the disease, and the most deadly. Dubbed "woolsorter's disease" in nineteenth-century England, it was spread by spores nestled in contaminated sheep wool. Before the introduction of sanitation and vaccination, workers in goat-hair mills were regularly exposed to high concentrations of anthrax spores; inexplicably, however, few suffered inhalational anthrax, perhaps because there weren't enough spores in the air or because the exposed workers developed some immunity. Between 1900 and 1976, there were only 18 reported cases of inhalational anthrax in the United States. Many of these cases were bizarre: a woman who played bongo drums made of infected skin, a construction worker who handled contaminated felt, gardeners whose infections were traced to contaminated bone meal fertilizer. In 1957, four people died of inhalational anthrax in the Arms Textile Company building in New Hampshire; all had worked with fibers from a single shipment of black goat hair from Pakistan, to be used in the manufacture of lining for men's suits. Ironically, the deaths took place during the trial of an anthrax vaccine that was quietly being conducted at the plant by the CDC. None of the workers who died had been immunized; after their deaths, the trial was discontinued and all the company's workers were required to take the vaccine.

America's current stockpile of anthrax vaccine, approved only to prevent the skin form of the disease, was derived from these 1957 studies and licensed in 1970 by the Food and Drug Administration. Reserved for military use, it can be released to civilians only with permission from the Defense Department. Weeks after the first anthrax cases in 2001, the vaccine received FDA approval for use in high-risk individuals, such as lab workers and hazardous material clean-up crews. Because the vaccine can have serious side effects and must be given in a complicated regimen, health officials have not recommended immunizing the entire population in the absence of a serious danger of widespread attack. Today, only one company—BioPort Corporation of Lansing, Michigan—manufactures the drug. But the company has not made the vaccine since 1998, having failed FDA inspection standards. Government and private researchers are working on new anthrax vaccines intended to be safer and easier to administer.

If released in a fine-particle mist, anthrax spores can ride air currents for 50 miles or more. Once a person becomes infected by inhaling the invisible spores, symptoms may not show up for weeks; in an accidental release of bioweapon spores in the Soviet Union in 1979, one patient didn't become ill until 46 days after infection. This delay means that, unless forewarned about a large-scale attack, health workers would mount a hopelessly tardy response at best. Once symptoms do appear, it's often too late. Antibiotics are most effective before a person becomes ill. Untreated, inhalational anthrax infections kill 80 percent of victims. In 1993, the Office of Technology Assessment, a research arm of Congress, estimated that if 220 pounds of aerosolized anthrax spores were released over Washington, D.C., between 130,000 and three million people would die—lacking a massive program to dispense prophylactic antibiotics—making such an attack potentially as lethal as a hydrogen bomb.

Perhaps a more loathsome agent of bioterror is smallpox. Caused by the orthopoxvirus variola, smallpox was officially declared eradicated from the face of the earth in 1980, after a tortuous 11-year World Health Organization (WHO) campaign, dubbed Target Zero, to accomplish just that goal. D. A. Henderson, the imposing physician and epidemiologist who was recruited from the CDC to lead the 1966–1977 effort—and who many believe should have received the Nobel Prize for his efforts—has referred to "innumerable instances in which the program balanced on a knife edge between success and disaster." In India alone, 120,000 field workers scoured the vast nation, tracking down every last case in every last village and vaccinating everyone in contact with the patient. To make sure smallpox patients stayed put until all their scabs had fallen off, field workers nailed shut the victims' doors and posted guards around the clock. In Ethiopia, one of the last bastions of smallpox, WHO crews sometimes trekked 100 miles to vaccinate nomads, risking death during the nation's civil war. In 1976, the very moment when Henderson was about to declare victory after the disease was finally snuffed out of that country, an outbreak surfaced in neighboring Somalia, requiring another year of searching for new cases, so that every last human contact could be vaccinated. Finally, in 1977, the last known naturally acquired infection occurred in a young Somali hospital cook, who survived.

Thus smallpox, which had killed 500 million people in the twentieth century alone, and infected one-tenth of all humans since the first agricultural settlements around 10,000 BC, became the first human disease to be eradicated as a naturally spread contagion. The next year, in a bizarre and tragic epilogue, a medical photographer in Birmingham, England, became infected and died, apparently when virus particles drifted through an air duct from a research lab one floor below. She infected her mother, who lived. As a standard public health precaution, the laboratory director was quarantined. Overcome with guilt and grief, he committed suicide.

In one of the bitter ironies in public health history, D. A. Henderson is today trying to help the world prepare for an intentional release of the virus he helped wipe out. Henderson founded the Johns Hopkins Center for Civilian Biodefense Strategies, and in the fall of 2001 became director of the Office of Public Health Preparedness within the U.S. Department of Health and Human Services, organizing a plan to defend against bioterrorism. Officially, the last remaining stocks of the smallpox virus are sequestered in two facilities sanctioned by the World Health Organization: the CDC in Atlanta and the State Center of Virology and Biotechnology (VECTOR) in Novosibirsk, Russia. But most observers, including Henderson, believe that other laboratories inside and outside Russia secretly harbor the virus. The WHO, which had approved destruction of the remaining official stocks, delayed that action until the end of 2002, to allow time for scientists to further study the virus's genetic structure and tricks for subverting the immune system, discoveries that may lead to antiviral therapies, vaccines, and rapid diagnostic and analytic tools in case of attack. The prospect of destroying the official stocks of smallpox virus had riven the research community. Many scientists insisted that the benefits of research outweigh the dangers of release, and that pretending the virus has not already proliferated is an illusion that serves no purpose. Others—notably Henderson—argued that there is little scientific insight that can't be gained using other orthopoxviruses, that the risks of accidental or intentional release eclipse any possible scientific advances, and that restricting the virus to outlaw states raises the moral bar against possible terrorist use. In late 2001, the Bush administration decided to retain U.S. smallpox stockpiles indefinitely, for research purposes.

If bioterrorists released smallpox virus, it would, by Henderson's reckoning, become a global calamity within six weeks. Before the disease was eradicated, smallpox spread naturally when an ill person released droplets from the mouth into the air. "When smallpox is epidemic," wrote a seventeenth-century observer, "entire villages are depopulated, markets ruined and the face of distress spread over the whole country." "Face" was meant literally. After the smallpox virus incubates for about 12 days in the body, victims suddenly feel weak, feverish, and achy; some become delirious. A rash appears, with especially dense lesions on the face, arms, and legs. The lesions are tense and deeply embedded in the skin. High fever, bloody sores, and seeping pustules give way to hemorrhaging, a sharp drop in blood pressure, and secondary infections leading to death. Should the victim live, the lesions will turn into scabs that eventually fall off, leaving depigmented scars. Though talk has abounded lately about suicide terrorists infected with smallpox, victims don't become contagious until the rash appears, by which time they would be far too ill to ramble about. Smallpox kills about 30 percent of its unvaccinated victims. Virgin-soil epidemics of smallpox, such as occurred in Native American tribes when European settlers came, have been known to kill as many as half their victims.

Smallpox radiates in ever-widening waves. Every silently infected person in the first swell of cases infects 10 to 15 more, who unless quarantined go on to infect 10 to 15 more. Given historical precedents, if the first generation of cases numbers 200 to 300, the next may be 2,000 to 3,000, and so on, if populations are not vaccinated and sick patients are not isolated. Henderson observed this steep rate of increase when smallpox raced through Yugoslavia in 1972. In that outbreak, a pilgrim returning from Mecca infected 11 family members and friends, who went on to spread the disease. In the four hospitals where the pilgrim stayed until he died, he infected 42 workers and other contacts. Not until four weeks after he became sick was the illness correctly diagnosed, by which time 150 people were already infected. In all, 175 people contracted smallpox and 35 died. So ominous was this outbreak in a nation that had not seen a case since 1946 that public health authorities acted with breathtaking speed. Within ten days, they vaccinated 20 million residents and quarantined some ten thousand under military guard in hotels and apartments. Neighboring countries sealed their borders. After nine weeks, the outbreak was halted.

Unlike with many diseases, immunity to smallpox begins to wane about ten years after vaccination. Here in the United States, where vaccinations ended in 1972, most residents have lost their protection, as have most people around the globe. There is less immunity in the world now than there ever has been in the history of humankind. The United States has only 7.5 to 15 million doses of vaccine—nowhere near enough to stem even a modest-sized epidemic. (The wide-ranging estimates of doses reflects the fact that administering the vaccine entails lightly puncturing the skin with a special bifurcated needle rather than giving a conventional injection, and unskilled vaccinators can waste the drug.) The most recent lot of smallpox vaccine was manufactured in 1982, and was produced in the traditional way—by scarifying the flanks and bellies of calves and harvesting infected lymph, a method that doesn't meet modern manufacturing standards for sterility. A new vaccine, in the works since 2000, will be grown in human cells suspended in large bioreactor tanks, and will be purer. After the 2001 terrorist attacks on the World Trade Center and the Pentagon, U.S. president George W. Bush decided to accelerate the expansion of the smallpox vaccine supply. That fall, with a rash of anthrax cases and the cloud of bioterrorism hanging over the country, Secretary of Health and Human Services Tommy Thompson negotiated a plan to produce enough additional smallpox vaccine to reach 286 million stockpiled doses by the end of 2002, enough to inoculate every American if necessary.

Until this ample supply is ready, the United States will depend on a federal emergency plan to respond to smallpox—a disease that last occurred in the country in 1949. According to this plan, if just one case of smallpox were confirmed, the patient would be immediately quarantined, and the CDC would dispatch vaccine from the government cache while alerting the FBI and the White House. Administered within a few days of exposure, smallpox vaccine can prevent or significantly reduce subsequent symptoms. CDC investigators would grill the victim's family about every step the patient had taken over the previous three weeks, and would ask for the addresses and phone numbers of every person who had close contact with the patient. State health officials would track down all those contacts—and *their* contacts—and vaccinate them. The plan is similar to the WHO's model during the smallpox eradication program—a plan specifically developed to make a little vaccine go a long way. Once there is enough vaccine for everybody, should all Americans be prophylactically immunized against smallpox? That may depend on whether smallpox materializes as a realistic threat. The smallpox vaccine produces adverse complications in approximately 1 in 13,000 people, ranging from rashes to lethal brain inflammations. Today, the risk of complications may be even higher because more people are living with immune systems weakened by conditions like cancer, HIV infection, and organ transplants. If the risk of dying from the vaccine is greater than the risk of getting the disease, mass immunization would make no sense.

Other potential biowarfare agents, though not all Category A, are nearly as frightening. Take plague. When its bacterium, *Yersinia pestis*, infects the lungs, causing a highly contagious form of the disease known as pneumonic plague, untreated patients quickly progress from fever and cough to respiratory failure, shock, and death; antibiotics are virtually useless if taken more than 24 hours after symptoms begin. Or consider the bacterial infection brucellosis, a.k.a. undulant fever, normally transmitted from cattle or goats or unpasteurized milk. A mere ten organisms can trigger symptoms from fever and sweats to weight loss and depression lasting sometimes longer than a year, coming and going in waves (thus its historic name). The sturdy bacterium that causes tularemia can remain alive for months in subfreezing temperatures; it brings on fever, prostration, weight loss, and pneumonia, and kills about a third of those not treated. Another potential bioagent, the rickettsial organism behind Q fever, kills 4 percent of victims and leaves the rest with throbbing headaches and eye pain that can last weeks. A third of its victims also develop hepatitis. Finally, viral hemorrhagic fevers, such as the mysterious Ebola or Marburg, begin with a high temperature, fatigue, and dizziness and can progress to bleeding under the skin, in internal organs, and from the mouth, eyes, or ears, leading to shock, coma, and death. There is no cure.

Although not living themselves, other potential weapons are derived from living things. These are especially terrifying. Botulinum toxins are the most lethal compounds known—15,000 times more toxic than the nerve agent VX and 100,000 times more toxic than the nerve agent sarin used in the 1995 Tokyo subway attack—and researchers estimate that as little as one gram of aerosolized botox could kill more than 1.5 million people. Within a day after exposure, victims experience blurred vision, difficulty talking and swallowing, and paralysis that creeps down from the shoulders and stills breathing. Ricin, a potent toxin easily extracted from castor beans, can be breathed in or ingested; if breathed in, it brings death from severe respiratory distress within days, and here too there is no treatment.

All told, of the 50 top bioweapon pathogens, only 13 have vaccines or treatments. In the short term, the advantage lies with the offense.

As frightening as this array of agents may be, imagine if their lethal capabilities were mixed and matched. By inserting genes from one organism into another, scientists may someday be able to design hybrid munitions that are more lethal, more sturdy, and perhaps even capable of eluding the immune system. "As a consequence, the quaint notion that you could list all of the bad pathogens that might be made into weapons and just forbid them and scan the world for them is ridiculous," says physician Tara O'Toole, director of the Johns Hopkins Center for Civilian Biodefense Strategies. "Because now all you have to do is click in the new gene, you get a new pathogen, you get a new weapon that ain't on the list." With this malign twist on molecular biology, writes D. A. Henderson, "the potential armamentarium is all but infinite."

Until recently, skeptics dismissed the potential of such malevolent biology, noting that virtually every time an organism was genetically manipulated, it ended up less, not more, virulent. Soviet experiments to hitch smallpox and Ebola viruses seemingly came to nought. But in 2001, an accidental discovery by Australian researchers gave the lie to the idea that designer diseases were pie-in-the-sky. While attempting to make a mouse contraceptive vaccine for pest control, the scientists had inserted into a mousepox virus a gene that makes large quantities of interleukin-4, a molecule produced naturally in mice and in humans. To their surprise, the designer virus crippled the ranks of the immune system that battle viral infection. As a result, nearly all the mice died from what normally is a mild infection. The new virus also resisted vaccination. Now scientists wonder whether terrorists, using similar techniques, could fashion human viruses that would wipe out the immune system. In 1998, Russian scientists reported that when they inserted genes from the harmless bacterium *Bacillus cereus* into the anthrax bacterium, *Bacillus anthracis*, they created a new form of anthrax that resisted both penicillin and vaccines.

Bioterrorism's potential doesn't end at human disease. Agroterrorism against crops and livestock could be just as devastating—which is why, shortly after the 2001 terrorist attacks, the Bush administration suddenly proposed spending tens of millions of dollars to hire more agricultural inspectors, a long-overdue action. U.S. agriculture has become dramatically less diverse genetically, making our food commodities as defenseless before an exotic pathogen as the Incas and Aztecs were before Europeans' smallpox. The centralization and globalization of our food supply leaves foods vulnerable anywhere along the chain, from farm to fork. About half of the American meat supply, for instance, is processed by three companies, meaning that livestock diseases will more readily spread because animals are concentrated in fewer places. A planned attack with foot-and-mouth disease—the highly transmissible infection that shook the British and European livestock industry in 2001—could theoretically wipe out a big chunk of American beef. Contaminating seed supplies or fields of soybeans with spores of soybean rust would reverberate globally, since the United States raises 50 percent of the world's soybean crop. From deliberate attacks with wheat rust and rice blast to sabotage of poultry with Newcastle disease or even the contamination of salad bars, the possibilities for mayhem are endless.

A Short History of Biowarfare

Until the autumn of 2001, skeptics dismissed talk of biological warfare as science fiction, the rhetoric of self-serving bureaucrats and professional paranoids. Some heard an echo from the Cold War 1950s, when the danger was said to be nuclear attack and schoolchildren memorized useless civil defense drills

But a look at the history of warfare—especially in the twentieth century—might reinforce our newfound fears. Biological weapons have been around since the beginning. Romans, Persians, and other ancients tossed carrion into wells and reservoirs to taint their adversaries' drinking water. In 1346, the three-year siege of the Black Sea port of Kaffa ended when attacking Tatars catapulted bodies of bubonic plague victims over the city walls; fleeing victims may have brought the Black Death to western Europe. In 1763, the commander in chief of the British forces in North America, preoccupied with a restive coalition of Indians on the Western frontier, hit on the idea of sending smallpox-infested blankets as gifts to the "disaffected tribes." During World War I, Central Powers spies infected Russian horses and mules on the Eastern front with glanders, which in turn infected soldiers.

"[P]estilences methodically prepared and deliberately launched upon man and beast . . . Blight to destroy crops, Anthrax to slay horses and cattle, Plague to poison not armies but whole districts—such are the lines along which military science is remorselessly advancing," Winston Churchill wrote in 1925. It was the year of the Geneva Protocol, the treaty that described chemical and "bacteriological" methods of warfare as "justly condemned by the general opinion of the civilized world."

Condemnation or no, biowarfare was about to enter its heyday. From 1932 to 1945, Japan carried out one of the most active biological warfare programs in history. The hub of Japan's program was Ping Fan, a small Manchurian village guarded by watchtowers, a moat, and a tall brick wall garlanded with high-voltage lines and barbed wire. There, directing the infamous Unit 731, Japanese army doctor Shiro Ishii explored the potential of plague, typhoid, paratyphoid A and B,

typhus, smallpox, tularemia, infectious jaundice, gas gangrene, tetanus, cholera, dysentery, glanders, scarlet fever, brucellosis, tickborne
encephalitis, hemorrhagic fever, whooping cough, diphtheria, pneumonia, meningitis, venereal diseases, tuberculosis, and salmonellosis.
At its peak, Unit 731 cultivated 660 pounds of plague bacteria each
month, as well as millions of fleas to carry the agent in airborne attacks. Researchers tied prisoners to stakes and detonated a shrapnel
bomb to infect them with the organism that causes gas gangrene. They
fed prisoners chocolates spiked with anthrax spores, biscuits laced with
plague bacteria, milk tainted with cholera vibrios, dumplings contaminated with typhoid. Army physicians practiced vivisection to observe
the disease process in real time. Ping Fan's researchers referred to their
test subjects as "logs." Predominantly Chinese citizens, these dehu-
manized victims included White Russians, Soviet prisoners, criminals,
and mental patients. During World War II, Japan reportedly attacked
at least 11 Chinese cities with biological agents. Western historians
estimate that, from 1932 to 1945, at least 10,000 prisoners died as a
result of experimental infection or execution following germ experi-
mentation. Calculations by Chinese and Japanese historians run higher;
they say at least 270,000 soldiers and civilians perished in this mon-
strous enterprise.

In 1947, the American military debriefed Shiro Ishii and other
leaders of the Japanese program and then cut an extraordinary secret
deal: immunity from prosecution if the Japanese would divulge to U.S.
interrogators the details of their biological experiments. Unbeknownst
to most of its citizens, America had launched its own biological warfare program, led by George W. Merck, president of the pharmaceutical company bearing his name. "Biological Warfare is, of course, 'dirty
business,'" Secretary of War Henry L. Stimson wrote in 1942 to President Franklin D. Roosevelt, "but . . . I think we must be prepared." A
special committee within the National Academy of Sciences supported
that view and concluded that militarizing microorganisms was "dis-
tinctly feasible." Indeed, committee members seemed inspired by the
prospect. "Meningococcal meningitis might be spread by spraying
meningococci in crowded quarters," they advised. "Typhoid could be

introduced by sabotage into water and milk supplies and by direct enemy action into reservoirs. . . . Botulinus toxin might be conveyed in lethal amounts through water supplies. . . . Plague could be introduced into any of the large cities or ports by releasing infected fleas or rats. . . . Diphtheria can be spread by dissemination of cultures in shelters, subways, street cars, motion picture theaters, factories, stores, etc., by surreptitiously smearing cultures on strap handles and other articles frequently touched."

In 1943, Camp Detrick (later Fort Detrick) in Frederick, Maryland, became the headquarters of U.S. germ warfare activities. Scientists experimented with anthrax, botulinum toxin, brucellosis, tularemia, psittacosis, plague, Venezuelan equine encephalitis, Q fever, cholera, dengue, shigellosis dysentery, glanders, Rocky Mountain spotted fever, and other human scourges, as well as the animal diseases fowl pest and rinderpest, and rice, potato, and cereal blights. As an insurance policy, U.S. researchers also worked on vaccines, toxoids, and other post-exposure treatments for the very organisms they were cultivating as weapons. By the end of the European war, the United States and Britain together assembled, though never released, a bomb that would rain anthrax spores over cities, leaving them potentially uninhabitable. The Americans had also developed a weapon that spread the livestock disease brucellosis—which, while highly infectious in people, is not as deadly as anthrax, and thus was considered to be a more "humane" weapon.

During the Cold War, the geopolitical rationale for biowarfare expanded, as did the U.S. arsenal. "There is no doubt that bacteriological warfare offers a unique psychological advantage," observed U.S. Army brigadier general William Creasy. "Man's dread of disease is universal. The mysteriousness and invisibility of bacteriological warfare agents, the knowledge that they strike via the simplest and most basic sources of man's security—food, drink, and the air he breathes—and a feeling of helplessness in dealing with the unknown, all add to the psychological potential." Creasy, who led the Army Chemical Corps's germ weapons program, fully understood the practical consequences of this high-stakes mind-game. "Biological warfare," he remarked in

1951, "is essentially public health and preventive medicine in reverse." American scientists searched for agents that were virulent, compact, reliable, and stable. To insure the safety of American troops advancing on poisoned terrain, researchers focused on pathogens that would disable victims but wouldn't spread person to person and wouldn't remain noxious in the battlefield. Above all, scientists wanted to create a dry, light bacterial or viral agent whose particles were minuscule— between one and five microns in diameter (a few dozen would line up across a single human hair), the better to reach the tiny alveoli in the lungs, where they could be absorbed into the bloodstream.

The secret U.S. program staged test runs in American cities using supposedly harmless bacterial stand-ins. Between 1949 and 1969, the military conducted 239 such open-air experiments over populated areas, from San Francisco to St. Louis to Minneapolis, to track how clouds of bacteria would drift and decay in the environment. In 1950, the navy staged mock attacks off the coast of San Francisco, releasing millions of particles of the bacterium *Serratia marcescens* toward the city. Researchers chose *Serratia* because it grows in pink colonies, making it easy to detect on culture plates set up through the metropolitan area. Unexpectedly, however, 11 patients with urinary tract infections triggered by *Serratia* turned up at local hospitals. One man died at a hospital that had never until then recorded such an infection; the true source of his fatal illness would remain classified for decades. Today, doctors know that *Serratia* does occasionally cause disease, especially in immunocompromised or otherwise debilitated persons. According to political scientist Leonard Cole, who has written two books on the covert operations, other individuals in the path of the military's practice exercises may also have sickened or died. "But we cannot know," he says, "and never will know."

The experiments continued. In 1964, the U.S. Army sprayed *Bacillus globigii*—as a spore-former, a good stand-in for anthrax—in Washington National Airport, proving that infected passengers could travel to more than 200 cities. In 1966, technicians dropped lightbulbs filled with simulants through ventilating grates and onto the tracks of the New York City subway, and calculated that a real attack using dried agents would have produced 12,000 cases of anthrax, 200,000 of tularemia, and 300,000 of Q fever; the wide range reflects how many—or how few—organisms are necessary to cause infection. During the Vietnam War, Seventh Day Adventists, who were conscientious objectors and served as noncombatants, volunteered to subject themselves to airborne tularemia, Q fever, and other agents, in a project dubbed "Operation Whitecoat." All told, between 1950 and 1969, the U.S. government sank more than $700 million into secret bioweapons research. By the end, scientists had transformed into weapons of war seven biological agents: *Bacillus anthracis,* which causes the human and livestock disease anthrax; *Clostridium botulinum,* a soil bacterium that produces botulinum toxin; *Francisella tularensis,* the bacterial agent of tularemia, or rabbit fever; various *Brucella* species of bacteria, which cause brucellosis, or undulant fever; the mosquito-borne virus causing Venezuelan equine encephalitis; *Staphylococcus* enterotoxin B, a bacterial source of food poisoning; and *Coxiella burnetii,* the rickettsial organism that causes highly infectious Q fever. Their methods remain classified to this day.

"We were fighting a fire, and it seemed necessary to risk getting dirty as well as burnt," recalled a leading Ft. Detrick scientist. "We resolved the ethical question just as other equally good men resolved the same question at Oak Ridge and Hanford and Chicago and Los Alamos." Though the scientists felt they were doing their patriotic duty, says Colonel Edward Eitzen, the current commander at the U.S. Army Medical Research Institute of Infectious Diseases (USAMRIID) facility at Ft. Detrick, "From our perspective, it seems like a perversion of science." But, Eitzen adds, "Whether this is right or wrong, a lot of what we know today about how to defend against biological agents is a direct offshoot of some of what was learned during that program."

In 1969, President Richard Nixon ordered the unilateral dismantling of the U.S. biological weapons program. Stockpiles were destroyed and the facilities for developing and producing them dismantled or converted to peaceful uses. "Mankind," Nixon stated, "already carries in its own hands too many of the seeds of its own destruction." Historians apparently never asked the president why he made that decision. They speculate that it may have sprung from a desire to forge a relationship with the USSR; from the realization that bioweapons are both unreliable in the field and risky to stockpile; and especially from fears that a strong biowarfare program in the United States would encourage Third World nations to embark on programs of their own, setting off a chain of one-upmanship that would ultimately boomerang on the U.S. Why, this last argument went, should the United States, the richest country in the world, with its massive arsenal of expensive and technically complex nuclear weapons, make war cheaper and easier for other nations? In 1972, the United States and the Soviet Union signed the Convention on the Prohibition of the Development, Production and Stockpiling of Bacteriological (Biological) and Toxin Weapons, and on their Destruction—more conveniently known as the Biological Weapons Convention, or BWC.

The very next year, the USSR embarked on the largest bioweapons buildup in its history. At its peak in the late 1980s, the program's 50-plus labs and testing sites would employ 65,000 researchers and technicians perfecting the science and art of germ warfare. The sprawling operation was camouflaged under the name Biopreparat, with an ostensible mission of developing civilian pharmaceuticals. In reality, Biopreparat was what has been dubbed a "toxic archipelago." Scientists toiled on 52 different agents that could be used as weapons, among them the organisms causing smallpox, anthrax, plague, Ebola and Marburg hemorrhagic fevers, yellow fever, tularemia, brucellosis, Q fever, botulinum toxin, and Venezuelan equine encephalitis. Genetic hybrids were whipped up from the most deadly ingredients.

It was 1979 when the United States first got an inkling of this enterprising activity. That year, intelligence experts picked up reports of an anthrax epidemic in the town of Sverdlovsk, an industrial center in the Ural mountains. From the start they suspected a biowarfare mishap, but could not prove it. Not until 1989, when a Soviet defector told British authorities about the Soviets' top-secret germ warfare program—tales of deadly bacteria nestled in warheads and bombs, impervious to heat and cold and drugs—did the West begin to fathom the depth of the Soviet program. In 1992, the deputy director of

Biopreparat—a defector who adopted the Westernized name of Ken Alibek—delivered even more shocking news. Biopreparat, he reported, had cooked up 2,000 strains of anthrax alone; a facility housing 7,000 employees worked on nothing but anthrax. According to Los Alamos National Laboratory molecular biologist Paul Jackson, "They've probably forgotten more than we'll ever know."

When the WHO declared smallpox eradicated in 1980, Soviet planners again discerned an advantage. "Where other governments saw a medical victory," Alibek wrote in *Biohazard,* "the Kremlin perceived a military opportunity." The USSR, he contends, produced 20 *tons* of smallpox virus each year—enough to kill the human population many times over—that could be mounted on intercontinental ballistic missiles and bombs. Protected by insulation and refrigeration, the viral payload would have produced an aerosolized cloud that, in theory, could have finished off any American survivors of a nuclear attack: a literal case of overkill.

Dissatisfied with the arsenal found in nature, Soviet scientists also attempted to fashion new organisms with enhanced properties for warfare. Though Western observers can still only guess at these accomplishments, accounts from former scientists who defected suggest that Soviet researchers did create bacterial and viral strains with higher virulence and stability than in naturally occurring strains, along with the capability to induce odd symptoms that would confuse diagnosis and treatment. They produced forms of plague bacteria that could secrete diphtheria toxin and that resisted antibiotics. They crafted viruses that genetically coded for bacterial toxins. They figured out how to make the fragile Marburg virus, which causes deadly hemorrhagic fever, rugged enough to be placed on a weapon. They fabricated "subtle agents" that could alter personality or make victims aggressive or sleepy. They even tried to recover influenza virus in corpses of victims from the 1918 pandemic, hoping to insert genes from the relict strain into currently circulating flu viruses. Alibek has asserted that Soviet Union researchers variously combined the viral agents of smallpox, Marburg, Ebola, Machupo, and Venezuelan equine encephalitis, though Western scientists are dubious. "Alibek has got a clear conflict of interest," says a U.S. government scientist. "He's in the United States now and has to make a living, and his area of expertise is Russian biological warfare. If that's not a threat, he's out of work." What's indisputably frightful is that, while Biopreparat sites have apparently stopped this research, today's Russian Ministry of Defense labs may still be carrying it on, under a cloak of secrecy that the West has not been able to penetrate.

As mind-boggling as these speculations are, the accidental 1979 anthrax outbreak at Sverdlovsk hinted at what real biowarfare would look like. Early on an April morning, at a military production plant for anthrax, a shift worker removed a clogged filter. The filter was part of the exhaust system of a drying machine that removed liquid from industrial scale cultures of anthrax spores. The worker forgot to replace the filter with a new one. No one knows how much time elapsed before someone noticed the error—perhaps a few hours. During that interval, an invisible plume of anthrax spores floated out of the plant and became windborne. The wind blew in a constant direction all day long, and the plume widened as it moved over nearby villages. People on the street, workers at a nearby ceramic plant, townsfolk sitting at their windows: all inhaled the spores. Two days later, dozens of victims appeared at hospitals gasping for breath, feverish, vomiting, their lips turning blue. Four days later, the victims began to die. Soviet authorities promptly confiscated medical records and officially blamed the outbreak on consumption of anthrax-contaminated black market meat. Sixty-eight died in the epidemic, among at least 79 infected. Sheep and cattle as far as 30 miles downwind also perished. According to Harvard University biologist Matthew Meselson, the anthrax spores that drifted on the wind could have weighed a total of anywhere from four milligrams to nearly a gram. Other Western scientists put the figure in the range of grams to kilograms. If an unintended accident could have such fatal results, what would happen in a well-stocked, deliberate attack? The Soviet Union, after all, at one time had 30 metric tons of anthrax ready to go.

The motive and the means for biological warfare have not been confined to the world's superpowers. In 1974, not long after the Soviet Union commenced its bioweapons buildup, Iraq—another signatory of the Biological Weapons Convention—began its own. When the Persian Gulf war erupted, Iraq was capable of launching missiles with biological payloads—but, for technical or political reasons, did not. Partly through United Nations inspections and through the admissions of Saddam Hussein's government, it is now known that Iraqi scientists worked on anthrax, botulinum toxin, cholera, plague, gas gangrene, *Salmonella*, ricin, staphylococcal enterotoxin, camelpox, cancer-causing molds called aflatoxins, rotavirus, and hemorrhagic conjunctivitis virus. During the Gulf War, the Hussein regime was alleged to have had aerial bombs and missile warheads packed with botulinum toxin, anthrax, and aflatoxins. All told, Iraq possessed at least half a million liters of various agents, enough to kill the world four times over. Many of Iraq's seed strains had been purchased between 1986 and 1991 from the American Type Culture Collection, a Maryland repository for government and academic researchers that has since tightened its export rules. While some observers feared that U.S. bomb attacks of Iraqi bioweapons sites might unleash epidemics, it never happened. For whatever reason, the organisms didn't survive or spread in the sunlight and air, nor have reports of local epidemics ever surfaced.

Iraqi officials have barred UN weapons inspectors from entering the country since December 1998. Intelligence experts suspect that Saddam Hussein's regime has rebuilt factories capable of producing chemical and biological agents, and may have resumed making weapons. They suspect Iraq can now produce a high grade of dry anthrax spores. Some observers fear that Iraq is working with camelpox, either as a way to create smallpox (the camelpox virus contains all the smallpox genes), or in order to manipulate camelpox so that it is virulent in people.

Iraq and Russia are not the only nations that the U.S. government suspects of harboring biological weapons. At least a dozen countries, including Iran, Libya, Syria, China, and North Korea, are believed to possess or to be trying to acquire such armaments. Western intelligence experts have sketchy evidence that North Korea's program may actually outstrip Iraq's. According to Alibek, Moscow State University for years trained scientists from Eastern Bloc states, Iran, Iraq, Syria, and Libya. After the breakup of the Soviet Union, Iran and presumably other nations tried to recruit Russian biologists for their own germ war enterprises. Whether Russian scientists—and Russian biological matériel—made

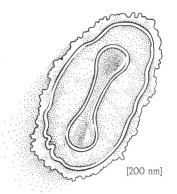

[200 nm]

Variola major
Agent of smallpox

their way to nations known to support terrorism is anybody's guess. Ironically, America's military supremacy may stoke the biowarfare ambitions of countries that could not defeat the United States using the approved forms of mass murder. After his nation's 1988 cease-fire with Iraq, Iranian president Hashami Rafsanjani described chemical and biological arms as the "poor man's atomic bombs."

The current debate over the bioterrorism threat hinges on whether the superpowers' technical expertise has leaked out. Some observers are convinced it hasn't. "It's not like Einstein and Madame Curie are rushing to join these terrorist organizations," says Amy Smithson, who directs the Chemical and Biological Weapons Nonproliferation Project at the Henry L. Stimson Center in Washington, D.C. According to Smithson, the technical hurdles that before September 2001 stood between terrorist groups and the highly efficient and effective dissemination of biological agents still exist. "Just because someone knows how to make something in a fermenter does not mean that they understand the intricate post-production steps that are required for technical dissemination. That's the reason why the former Soviet Union employed thousands and thousands of scientists: to get really good at the odious business of biological warfare."

Moreover, goes this argument, no rogue state sympathizer would be foolish enough to abet such terrorists. Even during the Gulf War,

Iraqi president Saddam Hussein observed certain boundaries. He had both biological and chemical weapons, and he didn't use them. Nor did the Soviet Union draw on its vast BW stockpiles during the Cold War. "It's not the moral restraints on the terrorists. It's the moral restraints on the leaders of nation-states," says Smithson. "It's pretty clear that Al Qaeda and its types have no problem with suicidal missions. But the purpose of a nation-state is to perpetuate itself. Cooperating with terrorists on finances or on training camps or places to hide and even on conventional weapons training is one thing. Cooperating with terrorists on chemical and especially biological agents is another thing entirely. That spells hell to pay."

But just because we have no tangible proof that terrorists possess biological weapons doesn't prove that they don't. Absence of evidence isn't evidence of absence—as recent history proves. Not until the West got the lowdown from Soviet defectors in the late 1980s and early 1990s did it begin to fathom the vastness of the Soviet BW enterprise. Even the accidental release of anthrax spores in 1979 from a Soviet military facility failed to tip off Western authorities. Likewise, the dimensions of the Iraqi biowarfare machine were unknown until UN inspectors started poking around after the Gulf War. "I am not reassured that we don't know about a teeny-weeny operation, possibly happening somewhere in the Mideast in the mountains of Afghanistan," says Johns Hopkins's Tara O'Toole. "And I don't think many people believe the CIA has tremendous expertise in these matters. It's very difficult to find these kinds of weapons before they're used, as everybody will admit." Here in the United States, it wasn't until after the fact that investigators found mountains of evidence that airline hijackings were about to take place—hardly an encouraging precedent for a mission to rout out unobtrusive miscreants with a stash of microbes. "What we saw on September eleventh is that terrorists are indeed quite well organized and sophisticated and capable of carrying out complex deeds requiring planning and determination," says O'Toole, "and that they're willing to cross the so-called barrier of moral repugnance and kill thousands of civilians without warning."

Methods of Madness

The prospect of bioterrorism may be closer than ever because undertaking it is easier than ever. "The main thing driving this is the trajectory of modern biological sciences: it is going straight up and fast as a rocket. There are a lot more people out there who know these basic biological techniques than there were atomic scientists in 1945," says O'Toole. Or as Nobel laureate Joshua Lederberg puts it, "The question . . . is: What levels of insanity do we have to prepare for?"

Technical expertise isn't some futuristic hypothesis—it's present fact. "There's no question in our mind," says O'Toole, "that organized terrorists could mount at least a small bioterrorist attack now." How big is "small"? In her estimate, "dozens or hundreds of people being infected in an indoor aerosol release, maybe more than one of those. . . . I agree that it's unlikely terrorists could create the kind of very efficient and accurate weapon that the U.S. had in the nineteen-sixties right now. I'm not so sure that will be impossible five years from now, given how technology is moving forward and simplifying the steps that you would need to isolate, harden, and disperse these organisms." Microbiologist Raymond Zilinskas, a senior scientist at the Monterey Institute of International Studies, agrees that a BioUnabomber is not a farfetched possibility. "I worry about the lone operative, the disgruntled or crazed scientist," he says. "That problem is going to grow, because as the population of microbiologists and biotechnologists grows, the absolute number of people that go bad would also grow."

Aspiring criminals can easily buy growth media and fermenters, agar and second-hand incubators, seed strains for common foodborne organisms. As investigators learned belatedly in the fall of 2001, laboratory samples of anthrax bacteria had for decades moved freely and without documentation among researchers and universities—samples ripe for theft. Moreover, the soil in certain locales is loaded with anthrax spores. With heat treatment and selective culture media, a competent microbiologist can easily tease out virulent strains. Converting the bacteria to spores merely requires adding certain chemicals or starving the organisms, and then culling out the spores. To spread microorganisms and toxins, motivated attackers can purchase off-the-shelf equipment from medical, agricultural, or industrial suppliers. Such experimental technologies as aerosolizers, currently being tested to spread vaccines to chicken flocks, will soon be available to farmers and anyone else with the cash. Asthma inhalers, crop-dusting equipment, photocopiers: all can spew out respirable particles of precise size and consistent quality.

The "Method" sections of articles in scientific journals are a gold mine for potential terrorists, describing precisely where the authors bought their materials and how they carried out their experiments. As Russian defector Ken Alibek testified before Congress in 2000: "Just by reading scientific literature published in Russia in the last few years, a biological weapons developer could learn techniques to genetically engineer vaccinia virus and then transfer the results to smallpox; to create antibiotic-resistant strains of anthrax, plague, and glanders; and to mass-produce the Marburg and Machupo viruses. Billions of dollars that the Soviet Union and Russia put into biotechnology research are available to anyone for the cost of a translator." More disquieting scientific "methods" are available in biological and chemical warfare cookbooks sold on the Web.

"An effective biological weapons program can be set up in a typical suburban basement, using basic high school or college lab equipment and materials easily ordered from catalogs," epidemiologist Mike Osterholm and journalist John Schwartz write in *Living Terrors.* They outline an easy if hypothetical attack on a crowded suburban mall. Using rapid-fasten Velcro strips, the villain attaches what appears to be a thermostat box to the wall. Inside is a microaerosolizer no larger than a pack of gum. Powered by a small store-bought camcorder battery, the unit transforms a few tablespoons of fluid—say, smallpox virus—into an invisible mist. Results: thousands would inhale the virus in the Muzak-filled ground zero, each infecting a dozen or so more in a multiplying geometry.

Though biological agents can be spread as liquids, dry powders disseminate far more easily. Particles one to five microns in diameter act like a gas, slowly settling from the air. But once released into the atmosphere, most biologic agents die or lose their virulence, their rate of decay contingent on a host of factors, including ultraviolet radiation, temperature, humidity, and pollution. Which means that the biggest remaining secrets from the U.S. and USSR biowarfare programs are how the agents were brewed and stored. The Cold War superpowers both developed techniques for suspending or dissolving optimal quantities of agents or toxins in special solutions containing preservatives, adjuvants, and antistatic chemicals. Every weaponized pathogen and toxin has its own formulation that prevents it from losing potency, clogging nozzles, clumping and falling to the ground, drying up in the atmosphere, or being killed by ultraviolet light.

Until the anthrax cases of 2001, the looming precedent for bioterrorists was the Aum Shinrikyo cult in Japan, which staged the 1995 sarin nerve-gas attack on the Tokyo subway system that killed 12 and injured upwards of 1,000. Before that headline assault, the group had 10 times tried and failed to sow disease by dispersing biological agents. Led by a former yoga teacher known as Shoko Asahara—who preached a mystical blend of Tibetan Buddhism, Hinduism, Christianity, Nostradamus, and pseudoscience—Aum Shinrikyo members came to believe that an apocalypse was just around the corner. To back up this prophecy and lend credence to their leader's divine revelations, cult members—many university-trained in the sciences—attempted to wreak havoc in cities and at U.S. military bases in Japan. Using a sprayer and fan, they repeatedly tried to disseminate anthrax from atop a cult-owned building in Tokyo. A specially fitted automobile spraying botulinum toxin through its exhaust toured downtown Tokyo during the 1993 wedding of Japan's Crown Prince. Whether because of the agent's inherent weakness or because of problems with misting devices, none of these forays succeeded. In frustration, the cult turned to a chemical for the subway attack. Sarin, after all, had a track record; in 1994, the cult's sarin attack in the town of Matsumoto had killed seven and injured 144, including several of the judges who were their targets. Just one month before the Matsumoto attack, Aum had mulled staging sarin attacks in the United States. Authorities suspect Aum also experimented with the agent of Q fever, and that a cult mission dubbed the "African Salvation Tour" traveled to Zaire in 1992 to obtain Ebola virus.

In the United States, the only successful large-scale bioterrorist attack was a decidedly low-tech incident of food poisoning in Oregon in September 1984. In this case, the culprits belonged to another cult—the Rajneeshees, led by a Nietzsche-inspired Indian guru named Bhagwan Shree Rajneesh—that had planned to take control of a rural county commission. They intended to swing the vote by sickening residents on the day of the election, having registered homeless people who would approve the cult slate. In their state-licensed clinical laboratory, called the Pythagoras Clinic, they grew a strain of *Salmonella typhimurium;* the organism was originally a control sample used to meet quality assurance standards for the lab. That September, cult members ran a field trial, pouring vials of the bacterium into salad dressings and coffee creamers at ten salad bar restaurants in The Dalles, a small town on the interstate. Within days, 751 people in the community became sick, though unreported cases among out-of-state travelers probably means there were many more victims. At the time, investigators considered bioterrorism a remote possibility. Despite the successful practice attack, the Rajneeshees decided against a repeat performance during the November elections. Not until a year later, when authorities were investigating the group for other criminal activities, did a cult member confess to the crime. The CDC delayed publishing details of the incident for 13 years, for fear of spawning copycat episodes.

Prior to 2001, the United States had seen only a handful of other small-scale attacks or planned assaults using biologicals. In 1996, a disgruntled lab employee at St. Paul Medical Center in Dallas infected 12 of her coworkers with *Shigella dysenteriae* type 2. (She had earlier practiced the trick with her estranged fiancé.) The organism, first described after a winter epidemic in an Eastern European prison camp during World War I, is now rare in developed countries. Using her supervisor's computer e-mail account, the employee anonymously invited her workmates to the staff break room, where she had laid out an enticing display of blueberry muffins and doughnuts—injected with

Shigella samples removed from the lab freezer. Everyone who partook got sick.

In 1994 and 1995, four members of the Minnesota Patriots Council, a band of antigovernment tax protesters, were convicted for conspiring to kill local and federal law enforcement agents with ricin, a deadly toxin made from the seeds of the castor bean plant. Two hundred times more potent than cyanide, and with no known antidote, ricin seems to carry a certain cachet among would-be conspirators, possibly because of the mistaken belief that it cannot be traced after it breaks down in the body. The Soviet KGB was said to favor ricin. In 1978, the Bulgarian secret police murdered dissident Georgi Markov in London by shooting a ricin-filled pellet into Markov's thigh from an airgun concealed in an umbrella. The Minnesota Patriots Council plot was somewhat less elegant. The group obtained its castor beans from a mail-order outlet called Maynard's Avenging Angel Supply. The strategy was to extract the ricin from the beans, blend it with a mixture of aloe vera and the solvent dimethyl sulfoxide (DMSO), and smear the concoction on doorknobs of the intended victims' homes or inside their shoes. A Council member called police before the plan was set in motion.

In what now seems like an innocent prelude, the United States in the late 1990s suffered an epidemic of bioterror bluffs. One of the first took place in April 1997, when the Washington, D.C., headquarters of the B'nai B'rith received a package leaking red fluid, inside which was a petri dish labeled *Anthracis Yersinia*: a fictional recombinant of anthrax and plague that doesn't exist in the natural world, but the invocation of which apparently terrified local authorities. Two city blocks were cordoned off, 109 people were quarantined, and 30 individuals underwent the humiliating strip-and-scrub of a hazardous materials decontamination. In the end, the suspicious substance was identified as red gelatin. That November, as part of a Clinton administration campaign to build public support for the imminent bombing of Iraq, Secretary of Defense William Cohen hoisted a five-pound sack of Domino sugar on a Sunday morning talk show, explaining that a comparable quantity of anthrax would kill half of Washington, D.C. In February

1998, Larry Wayne Harris, a loquacious "Christian Patriot" and well-known antigovernment character, was arrested in Las Vegas with what he himself touted as "military grade anthrax" (an ambiguous label) and which turned out to be a benign vaccine strain. A few years earlier, Harris had purchased plague bacteria from the same company that had sold Iraq starter strains for its biological cache. The media had a field day.

Soon after Harris's arrest came a deluge of anthrax threats—all hoaxes, many giddily covered by the press, which only fueled the fire. The crime du jour, anthrax hoaxes became so mundane that a California accountant accused of embezzlement and perjury actually called in a hoax to avoid a hearing in U.S. Bankruptcy Court. Chicago's Wrigley Building was shut down for six hours in response to an agent later found to be hot cocoa mix. From 1997 to early 2001, some 13,000 hapless building occupants had been ordered to strip and be hosed down as part of the misguided HAZMAT (for hazardous materials) emergency response to these ruses.

But hoaxes are just that—hoaxes—and do not necessarily reflect the real threat. Even before the attacks of 2001, an academic cottage industry had been weighing the relative bioterror dangers posed by white supremacists, Christian or Islamic fundamentalists, tax-protesting zealots, rogue dictators, disaffected employees in the life sciences, freelance extremists, wacky cults, and deranged individuals. In 1999, one scholar even warned of "pre-millennial tensions," a prophecy that seemed to spring not from a CIA briefing book but from the pages of *Glamour* or *Redbook*.

After October 4, 2001—when the first of a series of anthrax cases hit the news—the floodgates opened. In two weeks, the FBI received 2,500 ultimately unfounded reports of possible anthrax, from abortion clinics to libraries to government offices. Meanwhile, real anthrax spores were silently infecting postal workers, media employees, and other unsuspecting victims. Grappling with America's first organized bioterrorist attack, the public health system proved itself to be as unprepared as the sibyls had long warned.

The Home Front

No national emergency since the 1918–1919 flu pandemic has truly tested the U.S. health care system. During the pandemic, 28 percent of Americans became ill and 2.5 percent of those who were infected died. Just as with a broad-scale bioterrorist attack, doctors and nurses were in short supply, in part because they themselves had become sick. Medical students assumed physicians' duties. Druggists couldn't fill their orders and desperate customers quickly stripped pharmacy shelves of over-the-counter remedies. Gymnasiums, state armories, and parish halls were transformed into emergency hospitals. Caskets and burial plots ran out, and bodies were stranded in homes. At the Surgeon General's request, state and local health officials suspended public gatherings.

What impulses, good and bad, would be unleashed today if the United States were widely attacked with a bioweapon? Would people break into pharmacies to get drugs? Would a black market in antibiotics or vaccines flourish? Would fast food restaurants be commandeered to administer drivethrough prophylaxis for contagious diseases? Would citizens angrily second-guess official decisions about who gets rationed drugs? Would patients being cared for at home use the Internet to arrange for food and drugs? Would military or police authorities have to imprison contagious patients refusing to submit to quarantine, treatment, or vaccination? Would wild rumors circulate about the perpetrators of the attack?

Public reaction to the scattered but nerve-wracking cases of anthrax in 2001—with drug stores running out of the antibiotic Cipro, the worried well flooding hospital emergency rooms, and gas masks selling like hotcakes on the Internet—may be a pale presentiment of the country's response to a more widespread and murderous bioterrorist attack. Likewise, the fractures in the U.S. public health system revealed during the autumn of 2001 could become even more pronounced under a large-scale assault.

October 2001 did demonstrate that doctors' diagnostic reflexes are getting faster when faced with suspicious symptoms; like the first cases of West Nile virus in New York City, the initial cases of both inhalational and cutaneous anthrax were caught by astute physicians. But what if terrorism hadn't already been saturating the headlines? The first victims of an *unannounced* attack might drift in to emergency rooms, doctors' offices, and urgent care clinics. They would have vague flulike symptoms, since most infectious diseases begin with aches, fever, and chills—the immune system's first response to infectious invaders. At this stage of the outbreak, clues may be too subtle and too diffuse to pick up. Ideally, an alert doctor would raise a red flag if something were amiss and would call the local public health department. But if the early wave of patients could walk out, they would probably be sent home with a diagnosis of flu—as, tragically, was a Washington, D.C., postal worker in October 2001, who hours later died of inhalational anthrax.

To begin wrapping their minds around the practical challenges of a massive bioterrorist assault, health officials before the assaults of 2001 had begun staging so-called tabletop exercises, during which they rehearsed their decisions and actions around a table as a hypothetical attack unfolds. The results were revealing, but not especially reassuring. In 1998, when federal officials play-acted an outbreak of a fictional smallpox/Marburg hybrid virus along the Mexican American border, they discovered huge gaps in logistics and departmental turf wars. Hospitals sagged under the strain, federal quarantine laws failed, and an international political crisis exploded.

One of the biggest drills took place in May 2000 in Denver. Dubbed TOPOFF, because *top officials* from the federal government took part, the exercise simulated what would happen after a release of pneumonic plague at the city's center for performing arts. In this drill, public health officials, hospital employees, and political leaders stayed at their workplaces, as they would during a real incident. The drill unfolded according to a prewritten script that most of the participants did not know in advance. Participants learned about made-up events in this virtual attack—frightened citizens flooding emergency rooms, residents fleeing the city—from slips of paper handed to them over four days.

The drill kicked off when ten "patients"—in reality, healthy actors—began showing up at hospitals complaining of fever and cough. By the end of the first day, according to the script, there would be 783 cases of pneumonic plague and 123 deaths. Hospital staff called in sick, and antibiotics and ventilators were scarce. Though the script called for emergency "push packs" from the CDC's National Pharmaceutical Stockpile, Denver officials realized they didn't have enough people to unload the two airplanes full of drugs and supplies. In this city of one million, the official antibiotic distribution center would have been able to hand out drugs to only 140 people an hour—appallingly short of the goal of 100,000 people a day. In real conference calls during the drill, health officials took hours trying to agree on the proper dosage of antibiotics or deciding how close a potentially infected person would have had to stand to a victim in order to be considered "exposed." By day four, 3,700 people were infected and 950 were dead. Bodies piled up in morgues. In response to these scripted developments, city and state officials issued virtual orders commanding residents to remain in their homes and closing airports, bus stations, and train depots. With state borders closed, however, virtual food supplies ran out. Soon, cases emerged in other states, as well as in England and Japan.

In June 2001, in an exercise dubbed "Dark Winter," a mock National Security Council depicted by former senior government officials wrestled with a fictional smallpox outbreak that began after a release of the virus in shopping malls in Oklahoma City, Philadelphia, and Atlanta. Over a simulated time span of about two weeks, vaccines ran out, officials bickered over quarantine measures, and 6,000 Americans were dead or dying. By the finale, the imaginary epidemic had spread to 25 states and killed several million. When U.S. vice president Dick Cheney saw a video of the Dark Winter practice drill, shortly after the September terrorist attacks, he was so alarmed that he raised concerns about the smallpox vaccine supply that very day at a National Security Council meeting.

If the United States did face biowarfare, gas masks and private reserves of ciprofloxacin would be pretty much useless (gas masks because they would have to be donned immediately, Cipro because it doesn't work against many pathogens and it would engender antibiotic-resistant bacteria). Only a robust public health system—one that instantly registers aberrant syndromes and anomalies in infection rates, figures out the problem, and quickly intervenes—could actually curb the spread and devastation of the disease. In the ideal world, doctors and nurses would immediately recognize the unfamiliar symptoms of a wide spectrum of biowarfare agents; state and local health departments would continually collect data from hospitals about patients with suspicious pneumonias, meningitis, blood infections, diarrhea, botulism-like symptoms, rashes with fever, and fatal unexplained fevers—symptoms that, en masse, suggest deliberate infection; pharmacies would report spikes in over-the-counter drug sales; laboratories would perform rapid tests that would unmask an intentionally released pathogen; extra hospital beds and emergency supplies would be in place; a wide range of vaccines and antibiotics would be stockpiled; officials would know in advance precisely the decisions—about vaccinations, quarantines, travel restrictions, and so on—that they would make in a crisis.

Unfortunately, public health has long been considered a poor second cousin to curative medicine. Indeed, it's a bitter joke in the profession that a master's of public health is the only degree that *reduces* one's salary. In CDC labs, plastic sheeting protects equipment from leaky ceilings. So strapped for money are many local health departments, no one staffs the phones on weekends. Transforming this creaky system into one that can handle germ warfare is a herculean task. "One of the biggest lessons from Dark Winter was the clamor from the participants for more information," says Tara O'Toole. "What they wanted to know was: What is the scope of this attack? Is it one attack or multiple attacks? How many are sick? Where are they? Are things getting worse? Are things getting better? Everything they wanted to know was public health data. It wasn't information that could come from FBI or CIA or the military."

An advanced public health system, of course, pays double on its investment, because while girded for the possibility of bioterrorism, it is also ready for more common contingencies, such as a schoolyard meningitis outbreak or an urban hot spot of drug-resistant TB. "An emergency system that's dusted off and used only during a rare event isn't going to work," says Tara O'Toole. "These systems have to be part of our daily routine if they're going to operate during crisis." The rapid diagnostic techniques that make it possible to swiftly detect anthrax or plague could quickly diagnose pneumonia in nursing home residents, or antibiotic-resistant staph in premature newborns. The sensor technology that can identify biowarfare agents in the body or in the environment could be used to detect *Salmonella* in chicken or *E. coli* in ground beef. The distribution systems for delivering antibiotics and vaccines after a terrorist attack could deliver antivirals and vaccine during a flu pandemic. If all public health departments shared the same electronic surveillance architecture, they could instantly mesh data on any breaking epidemic, from *Cyclospora* to St. Louis encephalitis. A public health system that can handle a massive anthrax assault—or, even more horrifying, a return of smallpox—should be able to respond to just about anything.

But such a system is a long way off. Thirty percent of all U.S. hospitals are in the red and nearly 60 percent of academic medical centers can't meet their operating expenses. Hospitals are short on beds; thin or nonexistent profit margins and managed care demands for cost-cutting have forced them to send more patients home right after surgery and to operate near capacity all the time. Budget pressures have also forced doctors to order fewer laboratory cultures, instead just treating unidentified infections with broad-spectrum antibiotics—bypassing potential clues to a deliberately spread epidemic. Just-in-time management also means that hospitals have small supplies of drugs on the shelves; today, they often run short of drugs for common infections. A national shortage of nurses and pharmacists would compound problems in the event of a widespread attack. There's no slack in the system—no "surge capacity," to use public health jargon—should hundreds or thousands of people suddenly get sick. If hospitals run out of beds during an unremarkable flu season, they would be hopelessly besieged after a biological weapon release. In a survey of 30 hospitals in four states and Washington, D.C., published in 2001, none were prepared to handle large numbers of casualties caused by biological, chemical, or nuclear weapons; indeed, 26 hospitals reported that they could only handle 10 to 15 victims at once.

Back in 1951, when the CDC's Epidemic Intelligence Service got off the ground, there was a mystique about "shoe leather epidemiology." There still is—and some jobs, such as interviewing victims, still must be done face to face. But shoe leather can't keep up with today's complex and speeded-up world. Some state epidemiologists are tracking cases with pushpins on paper maps, instead of databases and software that can quickly link cases. "Many of my public health colleagues," CDC director Jeffrey Koplan testified before Congress, "are still working on technologies that involve paper and pen, telephones, while their kids are at home using the Web and Internet to order from Lands' End and Toys R Us." In the event of a large-scale attack, the dearth of real-time numbers will stymie health officials. "We're not going to be able to say how many cases there are, where they are, whether everybody came from the same hockey arena, or whether this is more than one attack," says Tara O'Toole. During the deadly 1918 flu pandemic, O'Toole adds, "The public health system lost credibility overnight because they couldn't say if the epidemic was waxing or waning."

This loss of credibility was glaring in the fall of 2001, when a steady drumbeat of anthrax news unnervingly sounded through every news cycle. Federal officials publicly disagreed about key questions, from the scope of contamination to the potency of anthrax spores delivered through the postal system to exactly how victims were exposed and what their treatment should be. Authorities warned of future attacks and in the same breath minimized an individual's risk of disease. CDC administrators blamed the FBI for withholding information critical to public health decisionmaking. To be fair, these officials were trying to keep up with an outbreak that didn't follow the script terrorism experts had written—the premises of which were partly based on old military experiments using research animals. Even so, no single government office or official seemed able or willing to coordinate the frantic, chaotic medical and law enforcement investigations. Can the new Office of Homeland Security, created in response to the September 11, 2001 attacks, pull together the political power and personal determination needed to cut through entrenched bureaucracies' deep suspicions of change and of each other?

One reason the United States was caught off guard in 2001 was a patchwork national policy on bioterrorism preparedness. Part of the problem lay on Capitol Hill: in the competition for a windfall of counterterrorism dollars in the late 1990s, public health advocates had been forced to jostle against sharp elbows from the military, law enforcement, and intelligence communities. In Congress, oversight for counterterrorism cuts across 11 Senate committees and 14 from the House. "This new mission of civilian biodefense has been dropped in upon an organizational landscape that is uncharted and basically unfriendly to the mission," observed Richard Falkenrath of Harvard's Kennedy School of Government. Civilian biodefense, he added, was "a homeless mission"—an irony, considering it was part of the EIS's original job.

Frustrated public health officials perceived budgetary brinkmanship instead of a reasoned analysis of the problem. National security leaders apparently didn't understand the consequences of biological attack. "The notion of it being an epidemic had escaped them," says Tara O'Toole. Peggy Hamburg, while assistant secretary in the Department of Health and Human Services during the Clinton administration, couched her arguments for public health improvements in terms of the nation's safety and security. Yet at the time, opponents often asked her why the United States needed to invest tens of millions of dollars to produce a vaccine for a disease—smallpox—that didn't exist. Hamburg concedes that some of her liberal colleagues were also nervous that the mission of public health had somehow gotten mixed up with the agendas of the military and of law enforcement. In September 2001, just before the first anthrax cases came to light, a General Accounting Office report presciently stated that the U.S. public health care system was fragmented and poorly trained to respond to germ attacks—as we all saw. "Turf wars and overlapping jurisdictions," the *New York Times* reported a few weeks later, "are hampering progress in the investigation of the anthrax outbreaks."

Back in the 1950s, Alexander Langmuir exploited the government's fear of Communist-inspired biowarfare to build a superb system of disease surveillance. But as historian Elizabeth Fee points out, "At the same time that funding for biological warfare research was increasing in the United States, funds for local health departments were cut sharply." Will the front lines again be shortchanged? In the event of a bioterrorist attack, as the GAO report asserted, "cities would probably be on their own for the first 24 to 72 hours." According to the Henry L. Stimson Center, a public policy research organization, the federal 2001 fiscal year budget for combating terrorism was $9.7 billion; of that, less than $100 million went to public health infrastructure and surveillance. Indeed, just in the month after the initial wave of anthrax attacks in 2001, many state health departments had consumed a full year's budget. No doubt, funding will grow and shift in the coming years. In late 2001, the Department of Health and Human Services requested billions of dollars to speed production of the smallpox vaccine, boost hospital preparedness, hire more epidemiologists, and tighten security at laboratories that handle bioterrorist agents. But all public health is local, and fortifying our defenses against germ warfare will above all require routing federal money to the immediate fields of battle.

Before the 2001 spate of anthrax cases, the small amount of bioterrorism preparedness money that did trickle down to state health departments improved readiness across the board. "It's such a dry desert in public health," says Tara O'Toole, "that the capacity to buy some basic equipment, to stand up some rudimentary epidemiological programs, to hire a couple of more people, will make a difference." Because of the federal government's previous investment in preparedness, state labs now have more Biosafety Level 3 facilities for dangerous pathogens that can cause disease through inhalation. And because the CDC had spent more than $8 million in 2000 to staff and supply a network of 81 public health labs to detect bioterrorist pathogens, the agency was relieved of some of its onerous laboratory caseload when the anthrax emergency struck.

Still, as health officials realized in 2001, the U.S. public health system desperately needs faster methods of detecting and treating deadly pathogens. Traditionally, lab technicians have had to grow organisms in culture media before making an ID—a process that sometimes takes days. Around the corner are portable devices that can quickly decipher the genetic material of a suspicious agent, compare its DNA to that of known strains, and even discern single nucleotides that may be giveaways for antibiotic resistance or genetic engineering. Scientists are experimenting with biosensors that connect living tissues to electronic chips that would trip an alarm. Researchers are also working on versatile treatments for terrorist-sown diseases. At the Pentagon's Defense Advanced Research Projects Agency (DARPA)—the toils of which led to the development of the Internet and of stealth aircraft—scientists are making antitoxins to neutralize the products of deadly bacteria, and topical patches to protect exposed victims against anthrax and other diseases. True to its reputation for unorthodoxy, DARPA is even breeding strains of bees that can track and follow the sources of airborne toxins. Meanwhile, the National Institutes of Health and the Defense Department are developing vaccines against every major weaponizable germ; theoretically, the vaccines could be administered prophylactically to hospital workers and police, and distributed to the general population after an attack to stop the disease from spreading.

Fingerpointing

Although September 2001 and its aftermath were highly publicized, that won't necessarily be so in future bioterrorism. The most insidious aspect of an intentionally planted epidemic is that it could be hard to distinguish from a natural disease outbreak. Public health officials would take notice any time lots of people suddenly became ill from a single disease agent or suffered pulmonary symptoms (suggesting an aerosolized microorganism or toxin); if large numbers of people became ill from an agent not previously seen in their geographic area; or if several deadly epidemics erupted simultaneously. But these same criteria also describe recent high-profile outbreaks of *natural* causation—outbreaks that themselves looked at first like bioterrorism but proved not to be.

The 1976 epidemic of a highly fatal respiratory infection at an American Legion convention in Philadelphia, for instance, had all the hallmarks of a deliberate chemical or toxin attack—especially since it took months to identify the bacterium that causes what we now know as Legionnaires' disease. When young Navajos started mysteriously drowning in their own lung fluid in 1993, not far from the U.S. National Laboratories at Los Alamos and at Sandia, rumors floated that the agent had escaped from the national labs or had been released as an act of genocide against the Navajo people—until it was discovered that the hantavirus later dubbed Sin Nombre virus comes from contact with rodent wastes. AIDS in Africa in the 1980s, dengue fever in Cuba in 1981, pneumonic plague in India in 1994, foot-and-mouth disease in hogs in Taiwan in 1997, Nipah virus among animals and humans in Malaysia and Singapore in 1998, West Nile virus in New York City in 1999, foot-and-mouth disease in Great Britain and Europe in 2001: all were at some point suspected of having been deliberately introduced.

Especially in nations where, unlike in the United States, infectious disease is endemic, bioterrorism is easy to suspect but hard to confirm. "Once an allegation is made, it is impossible to disprove it completely, since the nature of the weapon makes it almost invisible," writes historian John Ellis Van Courtland Moon. "If it is difficult to prove that it has ever been used, it is impossible to prove that it has not been used. Doubt is never totally exorcised."

And though the United States has been the victim of the most high-profile attack to date, it is not above suspicion as a possible culprit. Just a week before the September 2001 terrorist attacks, American media reported that the U.S. government had conducted secret research on biowarfare preparedness. The Pentagon had drawn up plans to reproduce a Russian genetically engineered strain of the anthrax bacterium in order to test the U.S. military anthrax vaccine, and had built a mock germ factory in Nevada from commercially available materials. Meanwhile, the Central Intelligence Agency had constructed a model of a Soviet germ bomblet that the agency feared was being sold on the international market.

Many experts believe these sub rosa experiments violated the spirit, if not the letter, of the 1972 Biological Weapons Convention. The treaty, at this writing ratified by 144 countries, states that signatory nations would "never in any circumstances develop, produce, stockpile, or otherwise acquire or retain" biological weapons. As prophylactic and defensive research, the two Pentagon programs were permissible under the BWC, says Barbara Hatch Rosenberg, a biologist and chairman of the Federation of American Scientists Working Group on Biological Weapons. (In October 2001, the Pentagon decided to continue its research on genetically engineered anthrax bacilli.) The CIA germ bomb, a potential delivery system for biological agents, was not.

The BWC is an agreement between nations, not international law. One of the provisions negotiated after 1972 compels signatory nations to report annually on their defensive research. "The U.S. government was incredibly shortsighted," says microbiologist Raymond Zilinskas, who served on a UN weapons inspections team in Iraq. "This kind of research is permissible under the Biological Weapons Convention if it's carried out in the open and reported as part of the confidence-building measures—neither of which was done. Here we are doing activities that, if we found out they were being done in Iraq or Iran or North Korea, we would probably immediately bomb the hell out of them."

"It makes it look like we're trying to get away with something," adds Rosenberg. "I don't believe the U.S. intends to develop or possess offensive biological weapons. But I think a lot of the world does, and I think this plays right into their hands."

Do researchers "need to create every monster bug in order to know how to defend against it?" asks Harvard University's Matthew

Meselson. As he sees it, "Anything that's done in dark secrecy is going to arouse suspicion. I am worried about the defensive work that's going on. I don't see any coordination or overall safety mechanism to make sure that we don't actually stimulate the very thing we dread. If we saw anybody else doing this, we would be very upset." Indeed, Meselson is convinced that the prospect of bioterrorism "is likely to depend not so much on the activities of lone misanthropes, hate groups, cults, or even minor states as on the policies and practices of the world's major superpowers." Our government, he says, should approach the problem "as though it were a species interest and not a parochial American interest."

To fortify the BWC, delegates have tried to tack on provisions requiring on-site visits to make sure nations are obeying the treaty and challenge inspections when nations are suspected of violations. In the summer of 2001, the Bush administration wiped out years of work on the protocol by being the only nation to reject the draft text, on the grounds that surprise inspections could threaten national security or reveal drug companies' commercial secrets. After September 11, it reaffirmed that stance. The decision, while roundly criticized, goes to a central dilemma: virtually all the agents and equipment needed to make bioweapons are also needed for legitimate medical and industrial purposes. Conversely, any operation that makes vaccines, antibiotics, feed supplements, or fermented beverages could be converted to making biological weapons. The razor-thin distinction is one of intent. Biowarfare materials are inherently "dual use," and thus difficult to police. According to Zilinskas, most secret biowarfare work is likely to remain hidden unless an accident tips off the world (as happened in 1979 in Sverdlovsk); a nation's defeat reveals information about its biowarfare program (as happened with Iraq after the Gulf War); or intelligence sources ferret out clues to a secret program.

Barbara Rosenberg fears that by pulling out of the protocols and conducting secret studies, the U.S. government gave license to rogue states to do their own dubious experiments—research that may eventually find its way to terrorists like the Al Qaeda organization. "I don't think terrorists can possibly launch an attack without support from a government that's carried out extensive work on biological weapons," she says. "In turning down the treaty, the U.S. has turned down one of the very few means we have for exerting pressure on foreign governments not to get into that."

While treaties and negotiations are under way to prevent an attack, the defense for either a natural or intentional epidemic is the same: a robust global public health surveillance system, internationally financed and managed. Notes Tara O'Toole, basic research done in the name of bioterrorism preparedness—a kind of BioApollo project, as ambitious as the mission to reach the moon—could benefit all nations. "Most of the mortality in the Third World in the coming decade is going to be from infectious disease," she says. "If we become so smart about the immune system and fighting off infectious diseases that we think we can handle just about anything that gets thrown at us—because we'll have the diagnostic capability and the ability to rapidly formulate a biological response—we're going to have spin-offs that make a momentous difference to the health of the Third World. That, in turn, should narrow the gap between the haves and the have-nots—which is at least part of the reason that these asymmetric weapons are attractive: as something to use against the great Satan of the American hegemon."

Time and again, history has shown that humankind's battle against disease can unite, at least temporarily, the enemies in shooting wars. Cease-fires have been brokered in Sudan, Sierra Leone, Angola, and other countries so that immunization days could be held. During El Salvador's civil war from 1985 to 1990, there were three annual ceasefires between government and guerrilla forces to provide every child in the country immunization and booster shots. These "days of tranquility," as they were called, give historian and political scientist Leonard Cole hope. "Those who insist that biological weapons are the weapons of the future must explain why they have not been weapons of the past. Why have these easy-to-make, easy-to-disseminate, inexpensive weapons almost never been used? The answer, at least in part, rests in reasons that inspired the days of tranquility." As he sees it, "A party that suspends fighting in order to eradicate disease one day is far less likely to spread it the next."

But after 2001, one must ask: What of groups that are willing to violate our deepest moral precepts—groups that have no political constituency, no goal of tranquility—groups that answer only to themselves?

Chapter 8

Think Locally, Act Globally

This disease not seldom attacks the rich, but it thrives among the poor. But by reason of our common humanity we are all, whether rich or poor, more nearly related here than we are apt to think. The members of the great human family are, in fact, bound by a thousand secret ties, of whose existence the world in general little dreams. And he that was never yet connected with his poorer neighbour, by deeds of charity or love, may one day find, when it is too late, that he is connected with him by a bond which may bring them both, at once, to a common grave.

— William Budd, *Typhoid Fever: Its Nature, Mode of Spreading, and Prevention* (1874)

Throughout this book, the reader will surely recognize a common theme: as modern life grows increasingly complex, humans and pathogens will more and more cross paths. And whatever the emerging infection threat, whether pandemic flu or inhalational anthrax, foodborne illness or insectborne disease, keen surveillance and rapid response are the answers. Without an exquisitely sensitive disease tracking system—a worldwide web of health care workers and laboratories and communication networks that can register the faintest signal of aberrant infection against background noise—novel diseases will continue to sneak up on us. Such a system must be built not just within the United States but around the world. Although this book has focused on the U.S., there are no national borders against infection.

For many reasons, affluent countries are reawakening to the fact that the growing disparity worldwide in health and wealth imperils all countries. As Laurie Garrett explains in *Betrayal of Trust,* "Global public health action on an ongoing basis would, if it truly existed, constitute disease prophylaxis for every locality, from rich nation to poor. New York City need not worry about its inability to stop plague at JFK Airport if India's infrastructure can do the job in Surat, preventing spread beyond that Gujarati city. And Tokyo need not fear Ebola if Congo's hospitals are sterile environs in which the virus cannot spread. Safety, then, is as much a local as international cause. In public health terms every city is a 'sister city' with every other metropolis on earth."

What would it take to achieve true global health? Though an in-depth discussion is beyond the scope of this book, the short answer is, again: sharp surveillance and response. Public health is essentially a rational, step-by-step process that applies to all outbreak investigations: define a problem, recognize it, find out what causes it, figure out how to control and prevent it. Health authorities need to be able to discern unusual clusters of disease, identify the cause of illness, track the geographic and demographic spread of an outbreak, estimate the magnitude of an infection, describe the natural history of a disease, list the factors behind an infection's emergence, and measure the effectiveness of interventions. Often this logical series of events begins with the hunch of a single physician who senses something amiss. But it also depends on observation posts that actively look for specific diseases and disease syndromes. It depends on laboratories that can run molecular fingerprints to quickly connect the dots between seemingly unrelated cases. And it depends on swift communications networks that alert far-flung health departments to potential outbreaks. In addition, governments must be willing to underwrite research on the biology and ecology of infectious diseases.

The United States has a long way to go to upgrade its own under-staffed, underequipped, and underfinanced federal, state, and local health departments. Improving public health globally is a much taller order of business. It will mean building sensitive monitoring systems in developing nations—and in many areas of Africa or India, where new infections are likely to originate and where they have historically gone undetected, it will mean first building the most basic health systems. Doctors and nurses around the world must be trained in the practical aspects of public health and be welcomed into an international network of health care colleagues. Independent laboratories and timely electronic reporting systems must operate free of government interference. Meanwhile, researchers must keep tabs on conditions such as altered habitats or large population movements that give rise to emerging infections. How close are we to approaching such a system, a kind of global Epidemic Intelligence Service akin to the CDC's élite corps of disease detectives? According to one U.S. government estimate, at least ten years away. Others say that goal is wildly optimistic.

No one knows the promise and perils of achieving global health better than William Foege, MD. Bill Foege (pronounced fay´-gy) is a towering figure in public health, and not just because of his lanky 6-foot, 7-inch frame. For the past 40 years, he has been at the center of action against many of the most important infections facing human-kind and has helped shape national and international counteroffensives. Revered in the profession, Foege has combined practical vision with passionate humanity. His career offers clues not only to where we've been but where we must now go.

Son of a Lutheran minister, inspired as a teenager by the writings of theologian and physician Albert Schweitzer, Foege embarked on a medical mission to Nigeria in the 1960s. His assignment began with a disease that was not emerging but disappearing: smallpox. Campaigns to eradicate a single disease underscore the value of nations working cooperatively rather than separately, and this was doubly true during the Cold War. Western nations had thrown their weight behind the international smallpox eradication effort, however, not out of altruism but financial self-interest; they were spending hundreds of millions of dollars yearly on vaccination, surveillance, and other measures to protect their own healthy populations from the disease. The eradication campaign had just gotten under way, and Foege was serving as a medical officer in a village in eastern Nigeria, living with his wife and young son in a hut with no running water or electricity. One of his first discussions with the village's elders came just a few weeks after his arrival; through an interpreter, the men swapped stories about the 1918 flu pandemic.

Health authorities believed that to eradicate smallpox in a region, 80 to 100 percent of the population had to be inoculated. In December 1966, vaccine was in short supply in Foege's jurisdiction; the shipment for mass inoculations wasn't due to arrive for a few months. When smallpox appeared in a remote village, Foege had to figure out how to hold back the epidemic. Spreading out maps of the district and working with two-way ham radios, he contacted missionaries and asked them to dispatch runners throughout the region to learn where else the disease had broken out. Using this information and analyzing family travels and market contacts, he made an educated guess about where the epidemic would jump. Foege's team vaccinated all residents in the affected villages and in villages where the disease would likely strike. Miraculously, four weeks later, though less than 50 percent of the population had been vaccinated, the outbreak screeched to a halt. "Surveillance/containment," as the method came to be known, revolutionized the perennially cash-starved smallpox eradication campaign by saving money and time. When the World Health Organization officially declared smallpox eradicated in 1980, it was in no small part because of Foege's calculated risk-taking.

In 1967, after the Nigerian civil war forced Foege out of the country, he joined the CDC. Over the next ten years, as he rose through the agency's ranks, a host of strange new infections came to light. Three of these were agonizing, often fatal fevers from Africa. In 1967, lab workers in Marburg, West Germany, began suffering flulike symptoms, progressing to rash, acute diarrhea, painfully peeling skin, and eventually uncontrolled internal bleeding coupled with heart and brain damage. Within a week or two, they were dead. Marburg fever, as it came to be known, had sprung from contact with the African green monkeys shipped from Uganda for cell cultures to help prepare vaccines. So frightening was the epidemic, it prompted U.S. government officials to construct the first maximum containment lab at the CDC. Two years later, at a mission station in Lassa, Nigeria, an American nurse died of a strange malady that began with high fever and throat ulcers and ended with gastrointestinal hemorrhage. The nurse who cared for her died; *her* nurse, who was stricken but survived, was transported back home to the United States, where a Yale virologist studying the new agent became infected and desperately ill, and a Yale technician working on a separate floor died after being infected with the escaped agent. Lassa fever underscored the virulence and transportability of a previously unknown African virus—but it wasn't the last shock from Africa. In 1976, a pair of hideous epidemics in Zaire and Sudan—marked by high fever, agonizing headache, vomiting and diarrhea, delirium, bleeding from every orifice, and in Zaire a 90 percent death rate—sent health officials scrambling. Ebola fever looked like a fictional Andromeda strain become real.

In 1977, Bill Foege became the CDC's director. Despite the scares from abroad—news of which, unlike today, stayed within the small public health community—infectious disease in America seemed about to turn the corner. Foege had intended to apply the epidemiology and surveillance methods honed in infectious disease research to what appeared to be America's next battle: chronic diseases fostered by lifestyle and the environment. But during his six years as director, new homegrown infections unreeled like a public health martial arts flick. Laboratory scientists had just identified the bacterium behind Legionnaires' disease. In 1980, toxic shock syndrome appeared in menstruating women using extra-absorbent tampons. In 1981, scientists identified the bacterium behind Lyme disease. In 1982, a McDonald's restaurant in Oregon was ground zero for the first known *E. coli* O157:H7 epidemic.

The term "emerging infections" hadn't yet been coined. "We called them new diseases, knowing they weren't," Foege says, "but they were new to us." He suspected that these infections wouldn't stay put, and his globalist convictions grew even stronger, in part because he had seen firsthand that infections new and old tend to become entrenched in poor and neglected parts of the world. Experts in other realms of public health harbored their own concerns about the potential for worldwide epidemics. Veterinarians had discovered weird new viruses in Africa and Central and South America that had jumped from animals to people. Military doctors had seen how hard it was to contain infections among troops stationed abroad. These practitioners at the front lines felt a mounting sense of unease that largely went ignored by the mainstream public health community.

In June 1981, the CDC's Morbidity and Mortality Weekly Report, the tip-sheet known as MMWR, carried the now famous report of a rare pneumonia in gay men, the early rumblings of the acquired immunodeficiency syndrome, or AIDS, pandemic. At first, Foege thought the outbreak would fade. But every week brought alarming new data. "It was a steamroller that just got bigger and bigger," he told one reporter. "There was nothing like it on this scale. You have to remember that there are not many things that are one hundred percent fatal beyond rabies. AIDS just did not follow the rules in any way." The Reagan administration refused to adequately fund research. To compensate, Foege quietly shifted monies from other programs within CDC. In 1983, after the Republican administration refused to put warning labels on aspirin bottles despite CDC findings that aspirin can cause potentially fatal Reye's syndrome in young children, Foege quit.

The Reagan White House's indifference to AIDS seared into Foege's brain the realization that every public health decision boils down to a political decision—and that to mold policy, public health experts must become adept in the ways of political influence. From 1984 to 1990, he created and led a partnership of United Nations agencies and nongovernmental organizations that raised worldwide immunization levels from 20 percent to 80 percent for six major childhood diseases. From 1987 to 1992, he served as executive director of the Carter Center in Alanta, which mounted attacks on two African scourges, Guinea worm and river blindness. Here, too, he observed how public health is powered by politics. "President Carter had access to heads of state," he says. "In a short meeting, you could end up getting a commitment that would bind the minister of health and the minister of finance and everyone else. This access allowed you to shortcut lots of things."

But political commitment was scarce as AIDS continued to sweep around the world. No disease would better illustrate emerging pathogens' global reach and the need for a coordinated response. Today, the United States has arrived at a kind of bleak truce with AIDS. About 40,000 Americans are annually infected with the human immunodeficiency virus, or HIV, which causes AIDS; each year, about 16,000 Americans die. Though the infection first struck white gay men disproportionately, it has since settled among minority populations, primarily blacks and Latinos. The last few years have also seen a resurgence in young gay men. By mid-2001, AIDS had killed more than 450,000 Americans and infected more than a million.

These numbers pale compared to the toll in the rest of the world. At the end of 2000, an estimated 36 million people worldwide were living with HIV, and nearly 22 million had died. Each year, 5.5 million people are newly infected with the virus: more than 10,000 a day. Ninety-five percent of victims live in developing nations. In sub-Saharan Africa—where AIDS is the leading cause of death, and where three-quarters of global AIDS deaths occur— the numbers prefigure social

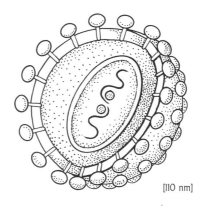

[110 nm]

Human Immunodeficiency Virus
Agent of AIDS

catastrophe. By 2010, life expectancy in Botswana will be 29; in Swaziland, 30; in South Africa, 36. And Africa is just the first great wave. By 2010, predicts a U.S. government National Intelligence Estimate, the number of HIV infections in Asia and the Pacific could easily surpass those in Africa. The Caribbean continues to suffer steep rates of infection. And in countries of the former Soviet Union, AIDS is spreading faster than anywhere else in the world, mostly through intravenous drug use. "There is nothing to suggest that HIV will plateau," noted *Nature* in 2001, "or that it will not reach 1 billion cases before 2050." A UN fact sheet starkly sums up the situation: "HIV will kill at least a third of the young men and women of countries where it has its firmest hold, and in some places up to two-thirds. Despite millennia of epidemics, wars and famine, never before in history have death rates of this magnitude been seen among young adults of both sexes and from all walks of life." By 2020, the AIDS virus will have caused more deaths than any disease outbreak in history—including the Black Death in fourteenth-century Europe, which mowed down one-third of the population.

Another disease erupted in the wake of the AIDS pandemic: tuberculosis, the nineteenth-century "white plague." The AIDS virus lowers resistance to infections and can ignite a latent case of TB into an active and contagious one. Today, more than two billion people—about a third of humanity—are infected with the TB bacillus, and each year two million die. The most menacing reservoir is Russia, where prisons—crowded, filthy, windowless—have become "pumps" for multidrug-resistant strains. What public health officials most fear are potentially untreatable TB strains that could emerge in Russia and Eastern Europe and race unchecked through the rest of the world, including the United States.

The great historic scourge malaria is another emerging infection, both because the mosquitoes that carry it resist pesticides and because the parasite itself is impervious to once-effective drugs. Over the last 25 years, the death rate of children with severe malaria in developing countries has not budged. Each year sees 300 to 500 million new malaria infections and one to three million deaths.

AIDS, TB, malaria: the top three infectious killers worldwide exemplify how poverty is a cauldron for disease. "The microbe is nothing; the terrain, everything," Pasteur wrote. Today, one in five people globally resides in poverty. Outside the privileged domain of American affluence, people contend daily with malnourishment, tainted drinking water, untreated waste, and ramshackle housing. Consider sub-Saharan Africa, where only half of the children are vaccinated against diphtheria, polio, and tetanus; sex education is rare; doctors or nurses are in short supply, as are laboratories to analyze blood samples; and many villages lie miles from the nearest passable road—a knot of problems that public health officials dryly refer to as inadequate "infrastructure." However it's described, it's a colossal barrier to disease prevention and treatment.

Poverty breeds compound infections and a vicious cycle of suffering. "It's possible to list causes of death in Africa," says Bill Foege. "But when a child dies of measles in West Africa, it in no way depicts the truth that that child was malnourished, that child also had malaria, probably had schistosomiasis and onchocerciasis and lymphatic filariasis and hookworms and roundworms and whipworms and repeated episodes of diarrhea. All of those things together allowed measles to be the final assault. We miss the point if we think that child had one disease. People in Africa with AIDS and tuberculosis—many of them were malnourished to start with. Many of them had malaria and all of the other parasites and STDs and so forth. So AIDS and tuberculosis come on top of that. When you look at combinations, you can't discount the role of poverty and illiteracy and fatalism. It's one big mess."

This mess has prompted physician and anthropologist Paul Farmer to question the very concept of "emerging" infections. Tuberculosis, he points out, has always had its tentacles in poor people. Hemorrhagic fevers and malaria have long been quotidian dangers in tropical countries. Deadly respiratory and diarrheal infections are a fact of life in Africa and much of Asia and Latin America. Even Ebola, when it ventures out of hiding, tends to strike people who rely on subsistence agriculture, and is amplified in rural hospitals that practice substandard medicine. "If certain populations have long been afflicted by these disorders," Farmer writes, "why are the diseases considered 'new' or 'emerging'? Is it simply because they have come to affect more visible—read, more 'valuable'—persons?"

Traditionally, leaders in the West felt they had to fix the "hard" problems in poor countries—problems such as lagging economic development and totalitarian regimes—before they could look at issues such as disease, which seemed merely a secondary complication. As a result, the status of global health today is utterly depressing. Yet paradoxical as it may seem, the pragmatic Bill Foege feels more optimistic than ever about its prospects. With newly forged links between governments, philanthropies, and private industry, the planets seem to be aligning in an auspicious formation.

And Foege is once again at the center of action, as the senior global health adviser to the Bill & Melinda Gates Foundation. By pouring billions of dollars into global health in a few short years, the Gates Foundation has become a powerful catalyst for change, creating a newfound sense of urgency about the international health crisis and shaming governments into action. "This is unprecedented," says Donald Hopkins, a Carter Center official and longtime public health leader, who has directed the worldwide campaign to eradicate Guinea worm. "Their flexibility and willingness to go where others haven't gone and get there fast, and the amount of money they have at their disposal, is transforming the field. Fifty years from now, maybe other organizations and official donors—governments and international agencies—will see that the way the Gates Foundation attacked these problems is the way they should move."

Not since the titanic era of Rockefeller, Carnegie, Morgan, and Ford has the United States witnessed such philanthropic largesse. The Microsoft founder's commitment to global health began in 1998, when he donated $100 million to create a children's vaccine program for respiratory and diarrheal infections, two of the biggest killers in developing countries. In 1999, the Gates Foundation donated $750 million to launch the Global Alliance for Vaccines and Immunizations, a coalition of international public health agencies, philanthropists, and drug companies that will work together to develop and deliver not only childhood vaccines for poor countries, but also vaccines for HIV and malaria and an improved vaccine for tuberculosis. In 2000, the foundation spent more than $1 billion on health projects around the world. The next year, it pledged $100 million to a United Nations fund to stop AIDS, malaria, and TB.

The foundation has channeled most of its contributions to improving access to existing and newly developing medical technologies. It pushes the use of simple tools such as vaccines, nutritional supplements, contraceptives, and diagnostic tests. In doing so, it straddles the old "vertical vs. horizontal" argument in public health over which is the best way to use scarce resources in poor countries: focusing on a specific disease or upgrading the public health system in general. Today, with a plethora of cheap and effective interventions, the answer is: both. "Deliver what you can deliver," Foege says, "and build the infrastructure as you're doing it."

The Gates' generosity has partly stimulated, partly complemented, other commitments to stopping disease. Even before the Gates Foundation began making its mark, media entrepreneur Ted Turner pledged $1 billion to a United Nations fund promoting global public health. In 2001, the UN General Assembly approved an extraordinary "Declaration of Commitment," outlining a worldwide campaign against AIDS. Late that same year, the World Health Organization teamed up with businessman George Soros to create a detailed $9.3 billion plan to reduce tuberculosis in countries hardest hit by the infection. The World Bank is exploring ways to convert debt relief to health grants, and has offered low-interest loans for HIV-related projects. The European Union and the G8 group of major industrialized nations are raising cash for health in developing nations. Coca-Cola, the largest private-sector employer in Africa, plans to use its fleet of trucks to distribute AIDS prevention and treatment literature and supplies to towns and villages where they are needed. Auto giant DaimlerChrysler will provide free AIDS drugs to help its South African employees and families combat the disease. Perhaps most important, current and former heads of state—from UN secretary-general Kofi Annan to former U.S. president Jimmy Carter—are leading the charge.

Even the traditional holdouts are starting to come around. Drug companies have often been averse to making vaccines or treatments for diseases in nations where people can't afford to pay (recent noble exceptions have included donated treatments for river blindness, lymphatic filariasis, and other diseases). These firms prefer to focus on ailments most prevalent in rich countries—as can be seen in the speedy commercialization of the Lyme disease vaccine, which has little use globally and is not even fully reliable here. "A corporation with stockholders can't stoke up a laboratory that will focus on Third World diseases, because it will go broke," Roy Vagelos, former CEO of Merck & Co., once observed. "That's a social problem and industry shouldn't be expected to solve it." But the bottom line may be starting to shift. Merck has donated $50 million for a pilot project to improve care and treatment in AIDS-ravaged Botswana. Pfizer has pledged $11 million for an AIDS care center in Uganda, and has offered to provide an AIDS treatment drug free to developing countries. And though manufacturers have long argued that the production of generic AIDS drugs violates their patents, international pharmaceutical companies in 2001 decided to allow South Africa to import and sell generic versions of expensive AIDS treatments; India and Brazil are making their own nonbrand versions of AIDS drugs. And the Gates foundation is working with drug manufacturers and developing countries to speed promising AIDS vaccines into clinical trials; while the companies will be expected to provide the vaccine at low cost in poor countries, in the United States they will be able to charge whatever the market can bear. "In the long run," says Barry Bloom, dean of the Harvard School of Public Health, "what's in it for the pharmaceutical industry is to have developing countries develop health care infrastructure for delivering any kind of drug."

"The landscape for international health has changed dramatically," adds Dr. Gro Harlem Brundtland, director-general of WHO. But can the enthusiasm and goodwill last? Some scientists claim that the massive new funding for AIDS, TB, and malaria has siphoned research money from other pressing diseases that are currently out of the lime-light. Public health underwriting, they suggest, is still a zero-sum game. Others say this massive new infusion of money could go to waste if the programs they support don't have clearcut and measurable targets in reducing disease. Yet others believe all this frenetic activity is just the new-scrubbed face of affluent nations' timeless self-interest—an up-dated version of colonial powers' desire to control tropical disease for their own motives. "Tie the needs of the poor with the fears of the rich," Foege says. "When the rich lose their fear, they are not willing to invest in the problems of the poor." Is the United States motivated to help Africa so that the continent will be a market for American prod-ucts or a producer of cheap goods for America? Or is the American stake primarily political? In 2001, a report prepared by the Interna-tional Crisis Group, an organization founded by former senator George Mitchell, warned that unless wealthy nations declare war on AIDS they will likely find themselves embroiled in conflicts fought by orphans in countries whose most productive adults died in the epi-demic. According to one official in the group, "It's like you've got an army waiting to be recruited for drugs, for crime, for militias."

"Ideally, I would like people to understand that there is no other way—that we're all so tied together globally that you have to think as a globalist," Foege says. "But if people don't believe that and for self-interest reasons do the right thing, I'm pleased to accept that."

Having started out as a medical missionary and ascended to the top of government agencies and having quit those for private founda-tions, Bill Foege has come to believe that each niche is only part of the solution. "The answer to health today and in the future is the develop-ment of coalitions," he says. "They turn out to be more important than the formal structure. The health leader is not necessarily the director of WHO—it's the person who could put together a proper coalition. We're seeing public–private approaches to global health problems that we have not seen in the past. If you're willing to think one hundred years in the future and ask, 'What will people wish we had done?' then you get a perspective of what global citizenship actually means."

At the end of 2001, the suddenly magnified threat of bioterrorism lent momentum to this process. As World Health Organization official J. W. Lee said of the newly unveiled global plan to combat tuberculosis, "Before September eleventh, many people thought that a nine billion dollar plan was impossible. But September eleventh proved money is not the issue. The issue is political will." Added Jim Yong Kim, a Harvard University physician long involved in fighting disease in impoverished nations, "The fear we feel now with anthrax is what people in the developing countries, TB-endemic areas, have been feeling every single day."

The price of not doing anything is incalculable—in part because we can't see the future. When smallpox was officially declared eradicated in 1980, no one knew that the AIDS pandemic was already afoot. People infected with HIV can't be vaccinated against smallpox because the vaccine decimates their weakened immune system. If smallpox hadn't been wiped out just in time, AIDS would have ensured that it never could be.

Discovering a method that hastened the end of smallpox was one of Bill Foege's great achievements. Yet when the WHO proudly announced that the disease was history, Foege felt no thrill. "When smallpox disappeared, I seemed to have none of the urge that a lot of people did to celebrate it. Not that I wasn't pleased that this had happened. But somehow my greatest high, my greatest reward, was at the beginning when I mentally came to the conclusion this could be done." Lately, Foege says, he has come to that conclusion about global health in general. Though it took 20 years for the AIDS pandemic to bring the world to its senses, Foege is convinced this heightened awareness won't soon dissipate. "I feel that we have just ended the incubation period for global response," Foege says. "Now the corner has been turned and people can never back off again. We're going to look back and think the AIDS problem is what finally caused us to see disease as the surrogate for an alien invasion, which required that the whole world figure out how to get together and respond."

Emerging infections are among the "thousand secret ties," in the words of Victorian physician William Budd, that bind the world into an organic whole. In rich nations such as the United States, the need for public health isn't always self-evident, because its product is invisible: when it works right, nothing happens. But here and abroad, public health systems must operate with the same constancy and efficiency as microbial evolution itself. It may take the worst emerging infection of all time to stir us into preparing for the next.

Selected Bibliography

Books

Barnaby, Wendy. *The Plague Makers: The Secret World of Biological Warfare.* London: Vision Paperbacks, 1997.

Beaty, Barry J. and William C. Marquardt, eds. *The Biology of Disease Vectors.* Niwot, CO: University Press of Colorado, 1996.

Beveridge, W. I. B. *Influenza: The Last Great Plague.* New York: Prodist, 1977.

Biddle, Wayne. *A Field Guide to Germs.* New York: Henry Holt & Co., 1995; New York: Anchor Books, 1996.

Bolduan, Charles Frederick. *Illustrious Contributors to Public Health.* New York: Department of Health, City of New York, 1936.

Brock, Thomas D. *Robert Koch: A Life in Medicine and Bacteriology.* Madison, WI: Science Tech Publishers, 1988.

Burnet, Sir Macfarlane and David O. White. *Natural history of infectious disease.* Cambridge: Cambridge University Press, 1972.

Butler, Richard. *The Greatest Threat: Iraq, Weapons of Mass Destruction, and the Crisis of Global Security.* New York: Public Affairs, 2000.

291

Christophers, Sir S. Rickard. *Aedes Aegypti (L.): The Yellow Fever Mosquito.* Cambridge: Cambridge University Press, 1960.

Cockburn, Aidan. *The Evolution and Eradication of Infectious Diseases.* Baltimore: The Johns Hopkins Press, 1963.

Cole, Leonard A. *The Eleventh Plague: The Politics of Biological and Chemical Warfare.* New York: W. H. Freeman and Company, 1997.

Collier, Richard. *The Plague of the Spanish Lady: The Influenza Pandemic of 1918–1919.* New York: Macmillan, 1974; London: Allison & Busby Ltd., 1996.

Craddock, Susan. *City of Plagues: Disease, Poverty, and Deviance in San Francisco.* Minneapolis: University of Minnesota Press, 2000.

Crosby, Alfred W. *America's Forgotten Pandemic: The Influenza of 1918.* Cambridge: Cambridge University Press, 1989.

Debré, Patrice. *Louis Pasteur.* Translated by Elborg Forster. Baltimore: The Johns Hopkins University Press, 1998.

Diamond, Jared. *Guns, Germs, and Steel: The Fates of Human Societies.* New York: W. W. Norton & Company, 1999.

Dubos, René. *Mirage of Health: Utopias, Progress and Biological Change.* New York: Harper & Row, 1959.

Ellner, Paul D. and Harold C. Neu. *Understanding Infectious Disease.* St. Louis: Mosby Year Book, 1992.

Endicott, Stephen and Edward Hagerman. *The United States and Biological Warfare: Secrets from the Early Cold War and Korea.* Bloomington, IN: Indiana University Press, 1998.

Etheridge, Elizabeth W. *Sentinel for Health: A History of the Centers for Disease Control.* Berkeley: University of California Press, 1992

Ewald, Paul W. *Evolution of Infectious Disease.* Oxford: Oxford University Press, 1994.

Farmer, Paul. *Infections and Inequalities: The Modern Plagues.* Berkeley: University of California Press, 1999.

Fenner, F. et al. *Smallpox and its Eradication.* Geneva: World Health Organization, 1988.

Fox, Nicols. *Spoiled: Why Our Food Is Making Us Sick and What We Can Do about It.* New York: Basic Books, 1997; New York: Penguin Books, 1998.

Futuyma, Douglas J. *Evolutionary Biology.* Sunderland, MA.: Sinauer Associates, Inc., 1998.

Garrett, Laurie. *Betrayal of Trust: The Collapse of Global Public Health.* New York: Hyperion, 2000.

Garrett, Laurie. *The Coming Plague: Newly Emerging Diseases in a World Out of Balance.* New York: Farrar, Straus and Giroux, 1994; New York: Penguin Books, 1995.

Goddard, Jerome. *Infectious Diseases and Arthropods.* Totowa, NJ: Humana Press, 2000.

Graves, Charles. *Invasion by Virus: can it happen again?* London: Icon Books, 1969.

Gregg, Charles T. *Plague! The Shocking Story of a Dread Disease in America Today.* New York: Charles Scribner's Sons, 1978.

Guillemin, Jeanne. *Anthrax: The Investigation of a Deadly Outbreak.* Berkeley: University of California Press, 1999.

Harris, Robert and Jeremy Paxman. *A Higher Form of Killing: The Secret Story of Gas and Germ Warfare.* London: Chatto and Windus, 1982.

Harris, Sheldon H. *Factories of Death: Japanese biological warfare 1932–45 and the American cover-up.* London: Routledge, 1994.

Henig, Robin Marantz. *A Dancing Matrix: How Science Confronts Emerging Viruses.* New York: Knopf, 1993; New York: Vintage Books, 1994.

Hope-Simpson, R. Edgar. *The Transmission of Epidemic Influenza.* New York: Plenum Press, 1992.

Hopkins, Donald R. *Princes and Peasants: Smallpox in History.* Chicago: The University of Chicago Press, 1983.

Karlen, Arno. *Man and Microbes: Disease and Plagues in History and Modern Times.* New York: G.P. Putnam's, 1995; New York: Touchstone, 1996.

Kilbourne, Edwin D. *Influenza.* New York: Plenum Medical Book Company, 1987.

Krause, Richard M., ed. *Emerging Infections.* San Diego: Academic Press, 1998.

Lederberg, Joshua, ed. *Biological Weapons: Limiting the Threat.* Cambridge, MA: The MIT Press, 1999.

Lederberg, Joshua, Robert E. Shope, and Stanley C. Oaks Jr., eds. *Emerging Infections: Microbial Threats to Health in the United States.* Washington, DC: National Academy Press, 1992.

Levy, Stuart B. *The Antibiotic Paradox: How Miracle Drugs Are Destroying the Miracle.* New York: Plenum Press, 1992.

Lilienfeld, Abraham M. *Times, Places, and Persons: Aspects of the History of Epidemiology.* Baltimore: The Johns Hopkins University Press, 1980.

Mangold, Tom and Jeff Goldberg. *Plague Wars: A True Story of Biological Warfare.* New York: St. Martin's Press, 1999.

Mattingly, P.F. *The Biology of Mosquito-Borne Disease.* London: George Allen and Unwin Ltd., 1969.

McKeown, Thomas. *The Origins of Human Disease.* Oxford: Blackwell, 1988.

McNeill, William H. *Plagues and Peoples.* New York: Doubleday, 1977; New York: Anchor Books, 1998.

Metchnikoff, Elie. *The Founders of Modern Medicine: Pasteur, Koch, Lister.* New York: Walden Publications, 1939.

Morse, Stephen S., ed. *Emerging Viruses.* New York: Oxford University Press, 1993.

Neustadt, Richard E. and Harvey V. Fineberg. *The Epidemic That Never Was: Policy-Making and the Swine Flu Scare.* New York: Vintage Books, 1983.

Oldstone, Michael B.A. *Viruses, Plagues, and History.* New York: Oxford University Press, 1998.

Osterholm, Michael T. and John Schwartz. *Living Terrors: What America Needs to Know to Survive the Coming Bioterrorist Catastrophe.* New York: Delacorte Press, 2000.

Patterson, K. David. *Pandemic Influenza 1700–1900: A Study in Historical Epidemiology.* Totowa, NJ: Rowman & Littlefield, 1986.

Peters, C.J. and Mark Olshaker. *Virus Hunter: Thirty Years of Battling Hot Viruses Around the World.* New York: Anchor Books, 1997, 1998.

Preston, Richard. *The Hot Zone.* New York: Anchor Books Doubleday, 1995.

Regis, Ed. *The Biology of Doom: The History of America's Secret Germ Warfare Project.* New York: Henry Holt and Company, 1999, 2000.

Rhodes, Richard. *Deadly Feasts: The "Prion" Controversy and the Public's Health.* New York: Touchstone, 1998.

Roberts, Brad, ed. *Biological Weapons: Weapons of the Future?* Washington, DC: The Center for Strategic Studies, 1993.

Rosen, George. *A History of Public Health.* New York: MD Publications, Inc., 1958.

Smith, Geddes. *Plague on Us.* New York: The Commonwealth Fund, 1941.

Speck, Reinhard S. *Bubonic Plague in San Francisco.* San Francisco: Five Trees Press, 1977.

Spink, Wesley W. *Infectious Diseases: Prevention and Treatment in the Nineteenth and Twentieth Centuries.* Kent, England: Wm. Dawson & Son Ltd., 1978.

Stanier, Roger Y. *The Microbial World,* 5th ed. Englewood Cliffs, NJ: Prentice-Hall, 1986.

Stern, Jessica. *The Ultimate Terrorists.* Cambridge, MA: Harvard University Press, 1999.

Theiler, Max and W. G. Downs. *The Arthropod-Borne Viruses of Vertebrates: An Account of the Rockefeller Foundation Virus Program 1951–1970.* New Haven: Yale University Press, 1973.

Tomes, Nancy. *The Gospel of Germs: Men, Women, and the Microbe in American Life.* Cambridge, MA: Harvard University Press, 1998.

Tucker, Jonathan, ed. *Toxic Terror: Assessing Terrorist Use of Chemical and Biological Weapons.* Cambridge, MA: MIT Press, 2000.

Vallery-Radot, René. *The Life of Pasteur.* Translated by R. L. Devonshire. Garden City: Doubleday, Doran & Company, 1928.

Waltner-Toews, David. *Food, Sex and Salmonella: The Risks of Environmental Intimacy.* Toronto: NC Press Limited, 1992.

Wills, Christopher. *Yellow Fever Black Goddess: The Coevolution of People and Plagues.* Reading, MA: Helix Books, 1996.

Zilinskas, Raymond A. *Biological Warfare: Modern Offense and Defense.* Boulder, CO: Lynne Rienner Publishers, Inc., 2000.

Zinsser, Hans. *Rats, Lice and History.* Boston: Little, Brown and Company, 1934.

Reports

Institute of Medicine. *Antimicrobial Resistance: Issues and Options.* Washington, DC: National Academy Press, 1998.

Institute of Medicine. *Assessment of Future Scientific Needs for Live Variola Virus.* Washington, DC: National Academy Press, 1999.

Institute of Medicine. *Orphans and Incentives: Developing Technologies to Address Emerging Infections.* Washington, DC: National Academy Press, 1997.

Institute of Medicine, National Research Council. *Chemical and Biological Terrorism: Research and Development to Improve Civilian Medical Response.* Washington, DC: National Academy Press, 1999.

Institute of Medicine, National Research Council. *Ensuring Safe Food: From Production to Consumption.* Washington, DC: National Academy Press, 1998.

National Research Council. *The Use of Drugs in Food Animals: Benefits and Risks.* Washington, DC: National Academy Press, 1999.

World Health Organization. *The Global Eradication of Smallpox: Final Report of the Global Commission for the Certification of Smallpox Eradication, Geneva, December 1979.* Geneva: World Health Organization, 1980.

World Health Organization. *Smallpox Eradication in India: Central Report to the International Assessment Commission on the Smallpox Eradication Programme in India.* New Delhi: World Health Organization, 1977.

Index

G

W

Y